Lecture Notes in Electrical Engineering

Volume 71

For further volumes:
http://www.springer.com/series/7818

Pramode K. Verma · Ling Wang

Voice over IP Networks

Quality of Service, Pricing and Security

 Springer

Prof. Dr. Pramode K. Verma
University of Oklahoma
4502 East 41st Street
Tulsa, OK 74135
USA
e-mail: pverma@ou.edu

Dr. Ling Wang
University of Oklahoma
4502 East 41st Street
Tulsa, OK 74135
USA
e-mail: ling.wang@ou.edu

ISSN 1876-1100

ISBN 978-3-642-26684-3

DOI: 10.1007/978-3-642-14330-4

e-ISSN 1876-1119

ISBN 978-3-642-14330-4 (eBook)

Cover design: eStudio Calamar, Berlin/Figueres

Printed on acid-free paper

Springer is part of Springer Science+Business Media (www.springer.com)

Preface

The growth of the Internet over the past decade, together with the promise of lower costs to the customer, has led to the rapid emergence of Voice over Internet Protocol (VoIP). This growth has been further fueled by the rapid penetration of the broadband all over the world. As a real-time application served by the Internet, VoIP faces many challenges such as availability, voice quality, and network security. This book addresses three important issues in VoIP networks: quality of service, pricing and security.

In addressing Quality of Service (QoS), this book introduces the concept of delay not exceeding an upper limit, termed the bounded delay (rather than the average delay), to measure the QoS in VoIP networks. Queuing models are introduced to address performance in terms of bounded delays. Closed form solutions, relating the impact of bounding delays on throughput of VoIP traffic, are developed. Traffic that exceeds the delay threshold is treated as lost throughput. The results addressed can be used in scaling resources in a VoIP network for different thresholds of acceptable delays. Both single and multiple switching points are addressed. The same notion and analysis are also applied on jitter, another important indicator of the VoIP QoS.

This book also develops a pricing model based on the QoS provided in VoIP networks. It presents the impact of the quality of VoIP service demanded by the customer on the transmission resources required by the network using an analytical approach. The price to be paid by the customer, as developed in this book, is based on the throughput meeting the performance criterion and the network transmission resources required. In particular, the impact of the QoS, as developed in this book, can be used in the design of VoIP networks in a way that would provide fairness to the user in terms of quality and price while optimizing the resources of the network at the same time.

This book also extends and applies the bounded-delay throughput analysis developed for VoIP networks in assessing the impact of risks constituted by a number of transportation channels, where the risk associated with each channel can be quantified by a known distribution. This discussion presents another application where the methodology developed in this book can be used successfully.

The overall economics of VoIP is addressed in Chap. 9. While the earlier chapters in the book have focused on consumption of bandwidth, this chapter discusses several factors that ultimately determine price to the user.

Security is a matter of concern in VoIP. For VoIP security, this book mainly focuses on signaling authentication. It presents a networking solution that incorporates network-based authentication as an inherent feature. The authentication feature that we propose introduces a range of flexibilities not available in the legacy network commonly known as the Public Switched Telephone Network (PSTN). Since most calls will likely terminate on the network of another service provider, the book also presents a mechanism that networks can use to authenticate each other. This mechanism affords the possibility of authentication across networks. Finally, this book explores areas for future research that can be built on the foundation of the research presented.

The book incorporates research conducted by the authors over a four-year period through 2009. VoIP is a rapidly growing application driven not only by lower costs but also by the availability of innovative applications that can be rapidly designed. Mobility can be considered another feature that is inherent in VoIP. Industry, academic researchers and graduate students can use this book. It is the authors' hope that it will offer a framework within which the users' interest (as perceived through QoS and pricing) and the service provider's interest in satisfying the customers' needs can be balanced without sacrificing profitability commensurate with investment.

Contents

Chapter 1
Introduction

Abstract This chapter introduces Voice over IP (VoIP) networking and the three major driving forces leading to its rapid growth. In comparison to the traditional calls through PSTN, VoIP offers several advantages; however, VoIP networks face many challenges as well, including quality of service and security. This chapter concludes by identifying the scope and content of this book in three aspects of VoIP networks: quality of service, pricing and security.

1.1 Motivation for VoIP Networks

Telephone systems have evolved over the last century. This evolution has comprised moving from analog to digital systems and from the circuit switching technology to packet switching systems. More recently, the potential benefits of converged networks as well as lower costs for the customer has led to the increasing use of Voice over IP (VoIP) [1].

Alexander Graham Bell is credited to be the first inventor to transmit voice electronically in 1876. Developments in telephony over the past 135 years have been nothing short of spectacular. The transmission and switching technologies have continuously evolved over this period. The most compelling changes in the transmission technology have led to the replacement of the copper wire by the glass fiber and, particularly, where mobility is desired, by wireless technologies. The switching technology has evolved from the circuit switching technology to packet switching technology. Packet switching affords the possibility of shared usage of a large-bandwidth channel as opposed to the use of slotted channels with a fixed bandwidth, traditionally of 64 kb/s, in circuit switching. Both the large bandwidth and the possibility of sharing can increase the efficiency of bandwidth utilization substantially. This will reduce transmission resources needed in the network.

Packet switching also constitutes the switching fabric of the Internet. As the number of Internet users has grown globally, the footprint of the Internet has

P. K. Verma and L. Wang, *Voice over IP Networks*,
Lecture Notes in Electrical Engineering, 71, DOI: 10.1007/978-3-642-14330-4_1,
© Springer-Verlag Berlin Heidelberg 2011

rapidly grown, making it almost as ubiquitous as the PSTN. In addition to ubiquity—which is a paramount requirement of communication—and the promise it gives of lower cost, there are two other factors that have fueled the growth of VoIP. Innovative applications can be easily created over the Internet and thus embedded as part of VoIP. Further, mobility is an inherent characteristic of the Internet because any terminal of the Internet retains its identity and uniqueness as it moves from place to place.

Rapid growth in the use of personal computers and the growing availability of broadband access to the Internet are the two other factors that have led to the growing popularity of VoIP. Thus, in summary, VoIP emerges as one of the most important services in the telecommunications industry driven by a large potential market, high availability of broadband access, and consumer willingness to accept the new technology to cut telephony costs while potentially reaping the benefits of a converged network.

As one example, Verizon, a major telecommunications service provider, stood out in 3Q'05 with 7.5% year-over-year decline in the number of its PSTN customers, which we suspect highlights competitive losses to VoIP services from multiple systems operators (or MSOs, e.g., Comcast, Time Warner, and Cablevision) in Verizon's Northeast markets [2]. Cisco also shipped over 980,000 residential VoIP gateways in 1Q 2005 equivalent to a 35% growth year-over-year).

Figure 1.1 shows the surge in the U.S VoIP residential subscribers in the first three quarters of 2005, indicating that the VoIP market will grow aggressively. In other words, it can be seen that carriers are losing more and more primary lines to VoIP substitutions.

According to the Telecommunications Industry Association, in Arlington, Va., in 2006, more than half of all the new private branch exchanges installed were IP-based and the number of residential VoIP subscribers is expected to rise 12-fold, to about 12 million by 2009, industry analysis project. By that time, the total U.S. revenue for business and residential VoIP products and services will be nearly $21 billion, up from $2.5 billion today, says Aaron Nutt, an analyst at Atlantic-ACM,

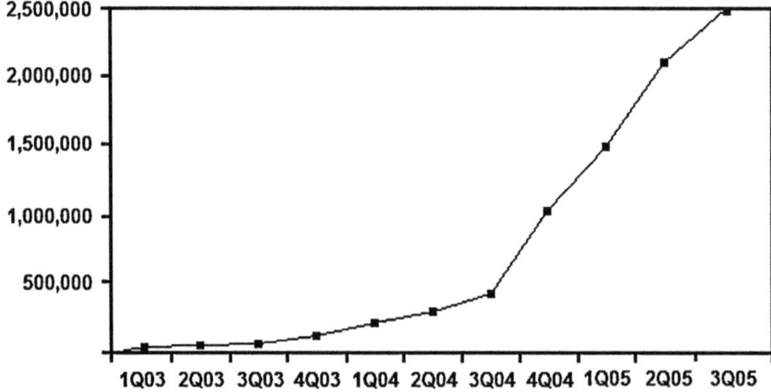

Fig. 1.1 U.S. VoIP residential subscribers [3]

a unit of Boston-based ACM Group Inc., which specializes in telecommunications consulting and market research [4].

The keen interest of the telecommunication industry in VoIP telephony is overwhelming, in spite of its relatively poor voice quality and lower overall availability of the Internet to support voice traffic, compared to the traditional circuit-switched telephony. Three major factors could explain such interest and the rapid development and deployment of the VoIP services. These are addressed as follows:

- Business drive
- Technology innovation
- Customer comfort

1.1.1 Business Drive

The primary factors that drive VoIP networks are business benefits. These are summarized as follows:

- *Consolidated voice and data network expenses*
  A single integrated telecommunications network with a common switching and transmission system is created to carry both the data traffic and the voice traffic. The integration of data and voice use the bandwidth and the equipment more efficiently. Voice packets share bandwidth among multiple logical connections alongside the data packets.
  In a traditional circuit-switched telephony system like the PSTN, in order to combine different 64 kbps channels into high-speed upper links for transmission, a substantial amount of equipment is needed. In the packet-switched IP telephony system, such as VoIP, in order to integrate the voice traffic and data traffic, the IP network utilizes statistical multiplexing. This consolidation represents substantial savings on capital equipment and operations costs [5].
- *Increased revenues from new services*
  VoIP networks not only support real-time voice communication, but also enable new services to be provided, such as instant messaging, unified messaging (voicemail to email), video conferencing and distributed games. These new services differentiate companies and service providers in their respective markets. Simultaneously, these new services encourage employees of the businesses using such services to improve productivity and, therefore, profit margins.
- *Flexible pricing structures*
  The establishments of VoIP networks push the service providers to build new pricing models. Since the IP network is shared by voice and data, and the bandwidth is dynamically allocated, the resource consumption of the network is not, generally speaking, measured in duration or distance as the circuit-switched telephony is. Dynamic allocation gives service providers the flexibility to meet the needs of their customers in ways that bring them the greatest benefits [5].

1.1.2 Technology Innovation

The second major reason for the rapid growth of VoIP networks is the development of technologies and innovations, which make VoIP networks feasible.

As we know, the main challenge of VoIP networks is that they cannot provide the same Quality of Service as PSTN does. However, the maturation of various codecs (voice coders and decoders) and high speed Digital Signal Processors, that perform voice packetization and compression, greatly improve the voice quality over the IP infrastructure.

Another aspect of technology innovations is the rapid emergence of new user applications and access to new communications devices, such as wireless devices, videophones, multimedia terminals, and personal digital assistants (PDAs). Upon demand by customers, service providers can much more easily enable new devices, offer larger volume of communications, and serve more subscribers in a packet switched environment. Furthermore, VoIP services have been aggressively marketed as a component of a compelling voice/data/video services bundle.

1.1.3 Customer Comfort

As we have mentioned above, access to new devices and advanced applications will satisfy customers and make their communication much more convenient and effective. VoIP has the potential to provide almost equal voice quality and multi-function service when compared to the traditional PSTN service, without increasing the price the customer pays. In VoIP networks, long-distance charges—including charges for international calls—can be, and usually are, transformed into a flat monthly fee or, in the case of the advertising-supported service providers, eliminated [6]. Indeed, the flat-fee charging model already exists for the Internet where most users pay a fixed monthly fee independent of the number of bits they send or receive. The amount of information transmitted or received is limited only by the access speed [7]. We might say that the potential savings on long-distance costs was one of the driving forces behind the migration to converged voice and data networks [5].

1.2 Comparison of PSTN and VoIP Networks

Over the last 100 plus years, voice services have been provided impressively through the PSTN. The PSTN core network uses 64 kb/s digital channels to provide dedicated end-to-end circuit connections in each direction. Voice terminals have both analog and digital access to the PSTN. Analog and digital telephone sets are generally connected to the central office with copper wire, called subscriber loops. Digital terminals use the ISDN protocols to access the network services.

In contrast to PSTN, VoIP technology digitizes voice and transmits data in frames over IP networks, where the reception of packets is not guaranteed because it relies on best-effort transport architecture. Even so, VoIP is rapidly establishing itself as an attractive technology for telephony with the maturation of related technologies and the recovery of the telecommunication industry.

PSTN has made very impressive achievement in terms of coverage, reliability, and ease of use. The number of current telephones lines is estimated to be 1 billion. People can hear the dial tone whenever they pick up the phone, and they expect to be connected to any destination in less than a minute. The availability of telephone service in such a plain old telephone system (POTS) is 99.999%, also referred to as a five-nine's reliability.

The Internet does not offer the same degree of reliability as the PSTN due to a variety of reasons. The complexity of multiple protocols, lack of standardization, multiplicity of equipment vendors and service providers, varying operating systems and network management systems all can cause lack of end-to-end interoperability. In addition, packet switched networks experience variable delays in the transmission process. In contrast, the PSTN doesn't suffer from variable delays, although it can experience blocking when all the available circuits are being used by other calls. The public Internet is collectively available only 61% of the time [8]. The best private data networks are available about 94% of the time, meaning that a user can be without the digital equivalent of dial tone about 22 days per year [9].

The PSTN-based Emergency-911 (E-911) services report the exact location of the telephone. Increasingly, this service is also required in VoIP. The location of a VoIP user is obtained by updating in the E-911 database. VoIP has not yet been regulated because IP-based telephony services are not regarded as traditional telephony services [10].

1.3 Major Challenges in VoIP Networks

1.3.1 Quality of Service

Quality of Service is one of the most important concerns in voice communications [11], determined by many factors, such as packet loss, speech coding options, delay, echo and jitter. The connection-oriented circuit-switched network provides each user with dedicated bandwidth for the duration of each call, which results in extremely low delay and jitter, and minimum disruption due to "noise" on the connections. High quality provided by the PSTN and private PBX-based networks drives telephone users to expect high QoS of the VoIP [12, 13].

As we know, VoIP uses different codecs, and codecs affect the quality of voice in a significant way, so it is especially important to measure the quality of a voice call in a standardized manner. One such measure is through the Mean Opinion

Table 1.1 ACR or MOS Scale

MOS	Opinion	Description
5	Excellent	Greater than toll quality
4	Good	Toll quality
3	Fair	Mobile phone quality
2	Poor	–
1	Bad	–

Score (MOS) [14]. According to ITU-T, opinion rating is generally used to assess subjective quality, which is the measurement based on a large number of users' perception of service quality under various conditions. One of the most frequently used opinion scale is shown in Table 1.1 [14]:

VOIP systems should achieve a level of quality near the toll quality, have a low delay, and good resilience to packet loss, but a lower cost [15].

1.3.2 Pricing

An important issue in designing pricing policies for today's networks is to balance the trade-off between traffic engineering and economic efficiency [16, 17]. A recent work [18] has addressed the impact of multiple hops (or switches between the ingress and egress switching nodes) on the grade of service offered by a circuit switched telephone network. A similar approach is adopted in this book in the context of packet switching. Accordingly, the grade of service is replaced by a threshold delay which is an appropriate measure for perceived quality of service in a packet switched network. In a circuit switched network, an incomplete call is lost and does not generate any revenue. In the packet switched networks, there are no calls that are lost as such; however, some of the packets may suffer delays above the acceptable delay bound and are, similarly, not considered to constitute effective throughput. Just as a caller in a circuit switched network does not pay for an incomplete call, the VoIP caller over a packet switched network in our construct should not have to pay for packets that suffer an unacceptable level of delay.

1.3.3 Security

Voice over IP applications are generally designed to function over the global Internet, although such solutions can be offered over private IP networks, such as enterprise networks. Instances of violations of security over the Internet are common occurrences that affect individuals, businesses as well as government operations. Voice over IP has not yet suffered many security violations, but the potential for attacks on security is truly large [19]. Possibly the mass of end points connected to VoIP today is below the threshold that would attract miscreants.

However, faking the identity of the caller can be easily accomplished using the Internet. This can result in the unsuspecting receiver of a call passing sensitive information to a caller who is pretending to be someone else such as an authorized employee of a bank where the receiver of the call maintains an account. Additionally, the negative impact of SPAM over Internet Telephony can be easily comprehended.

1.4 Scope of this Book

Delay is the most important parameter in voice communications due its effect on interactivity for real-time applications. The average delay has been traditionally used as a key measure of network performance. However, from a user's perspective, the upper bound of delay which controls jitter in a significant manner is far more important. The notion of delay not exceeding an upper limit, termed the bounded delay (rather than the average delay), is introduced as a measure of the quality of service in VoIP networks. Closed form solutions relating the impact of bounding delays to specified levels on throughput of VoIP traffic are developed. The analytical approach adopted can be used in scaling resources in the VoIP network for different thresholds of bounded delays. Similar results are presented for controlling specified bounds on jitter in a multi-hop network.

Analytical results are developed relating the quality of VoIP service to transmission resources required by the network. The price to be paid by the customer is based on the throughput meeting this criterion and the network transmission resources required. The results provide fairness to the user in terms of quality of service and price, while optimizing the resources of the network at the same time.

The analytical results developed in conjunction with VoIP are further used in an entirely different application. Using these results, Risk Analysis models are developed which can assess the cumulative impact of risk constituted by a number of channels, where the risk associated with each channel can be quantified by a known distribution.

Furthermore, a mechanism using VoIP networks, that can mutually authenticate each other in order to afford authentication across networks, is presented in the book. This proposed authentication plan introduces a range of flexibilities not available in the PSTN.

1.5 Organization of the Book

This book consists of three main parts. The first part (Chaps. 2–6) deals with the Voice over IP networks. Chapter 2 introduces VoIP networks, including voice quality, transport, Network QoS, call signaling and security. Chapter 3 provides an overview of the applicable queuing theory and presents an analytical model for delay-throughput analysis. Chapters 4 and 5 provide closed form solutions relating the impact of bounding delays on throughput for M/M/1 and M/D/1 models, respectively. Detailed analytical solutions and simulation results are presented as well. Chapter 6 considers the impact of bounded jitter on resource consumption in a multi-hop network.

The second part of the book (Chap. 7) addresses the impact of Quality of Service on resource consumption and proposes a pricing plan whereby the customers are fairly charged for various levels of QoS. In particular, the impact of Quality of Service presented can be used in the design of VoIP networks in a way that would provide fairness to the user in terms of quality of service and price, while optimizing the resources of the network at the same time.

The third part of the book is comprised of Chaps. 8, 9, and 10. Chapter 8 focuses on using the techniques developed in this book in an entirely different context—in assessing the cumulative impact of inhomogeneous channels on end-to-end risk in a transportation system. Chapter 9 addresses the economics of VoIP systems from an overall perspective. Chapter 10 reviews the potential attacks and current countermeasures on VoIP networks, including related algorithms and techniques. It presents a new authentication scheme that incorporates network-based authentication as an inherent feature, which supports a range of flexibilities not available in the PSTN. A mechanism that uses authentication across disparate networks is also presented. Chapter 11 presents conclusions and directions for future research.

Chapter 2
Voice over Internet Protocol

Abstract This chapter presents an overview of the architecture and protocols involved in implementing VoIP networks. After the overview, the chapter discusses the various factors that affect a high quality VoIP call. Furthermore, the chapter introduces various codecs and the engineering tradeoffs between delay and bandwidth. Finally, the chapter gives a detailed explanation of the currently widely used VoIP call signaling protocol, the Session Initiation Protocol or SIP.

2.1 VoIP Architecture

2.1.1 VoIP System

VoIP calls can take place between phone-to-phone, PC-to-PC, and phone-to-PC. The VoIP system configuration [20], shown in Fig. 2.1, is a representative scenario. In the PC-to-PC call, as an example, once the media path is established, the analog signal is sampled at 8 kHz or another frequency depending upon the codec. These samples are then encoded in an appropriate binary format. The encoded samples are put into UDP packets of different sizes and sent over the Internet. The reverse process takes place at the receiver PC: the speech samples are extracted from the packet, processed, and then put into the play-out buffer as the analog speech signal.

2.1.2 VoIP Protocol Structure

Since the 1990s, the dominant commercial architecture uses the Internet protocol suite TCP/IP, whereas VoIP uses RTP/UDP/IP. Figure 2.2 gives the complete communication network architecture.

P. K. Verma and L. Wang, *Voice over IP Networks*,
Lecture Notes in Electrical Engineering, 71, DOI: 10.1007/978-3-642-14330-4_2,
© Springer-Verlag Berlin Heidelberg 2011

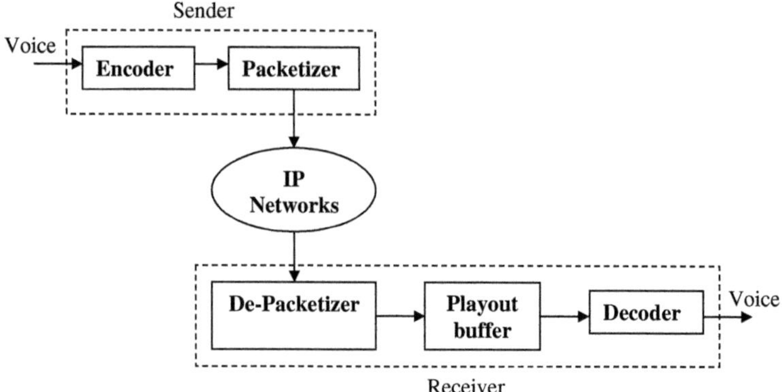

Fig. 2.1 Conceptual diagram of a VoIP network

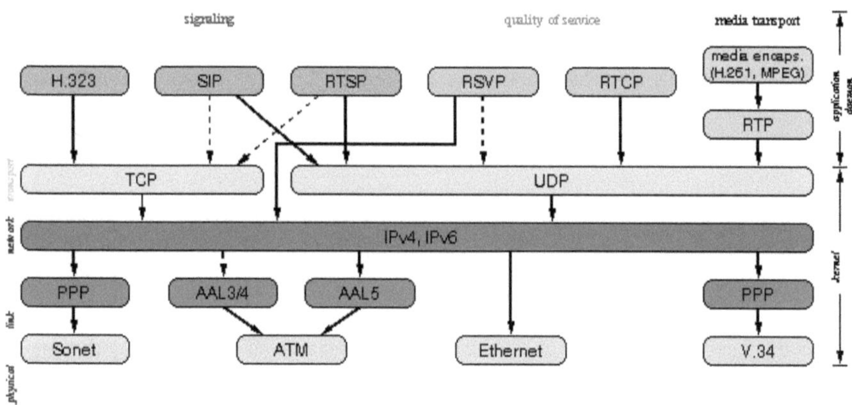

Fig. 2.2 Internet protocol stack [21]

As we know, the Internet Protocol (IP) deals only with the connectionless delivery of the packets, which is based on a best-effort service. Transmission Control Protocol (TCP) is a reliable connection oriented control protocol above IP. The TCP has the following characteristics. The TCP is:

- *Reliable*

Each transmission of data is acknowledged by the receiver, and retransmission is needed to ensure packet receipt in case of packet loss or error in the packet.

- *Connection oriented*

A virtual connection is established before any user data is transferred.

- *Full Duplex*

The transmission is provided in both directions.

- *Rate Adjustment*

The transmission rate increases when no congestion is detected; the transmission rate reduces quickly when the sender does not receive positive acknowledgments from the receiver within a stipulated timeframe.

Despite these features, the TCP/IP is not suitable for real-time communications, such as speech, because the acknowledgment/retransmission feature would lead to excessive delays [21].

In contrast to TCP, User Datagram Protocol (UDP) is classified as unreliable connectionless protocol, which does not provide sequencing and acknowledgement. Without flow control and error recovery, UDP simply sends and receives IP traffic between users in an Internet.

The Real-Time Protocol (RTP), used in conjunction with UDP, provides end-to-end network transport functions for applications transmitting real-time data, such as audio and video, over unicast and multicast network services [22]. RTP standardizes the packet format by including the sequence numbers and time-stamps, which is convenient to multimedia applications. It should be emphasized that RTP in itself does not provide any mechanism to ensure timely delivery of data or provide other quality of service guarantees [23]. Indeed, RTP encapsulation can only be seen at the end user location, and is not distinguishable from IP packets without RTP at the intermediary routers.

A companion protocol RTCP does support the features as follows:

- Monitors the link
- Separates packets sent on a different port number
- Exchanges information about losses and delays between the end systems
- Sends packets in intervals based on number of end systems and available bandwidth

However, a continuous stream of RTP/UDP/IP packets is offered in most VoIP applications as shown in Fig. 2.3.

As far as VoIP call signaling protocols are concerned, there are peer-to-peer control-signaling protocols such as H.323 protocol suites [24] and SIP [25],

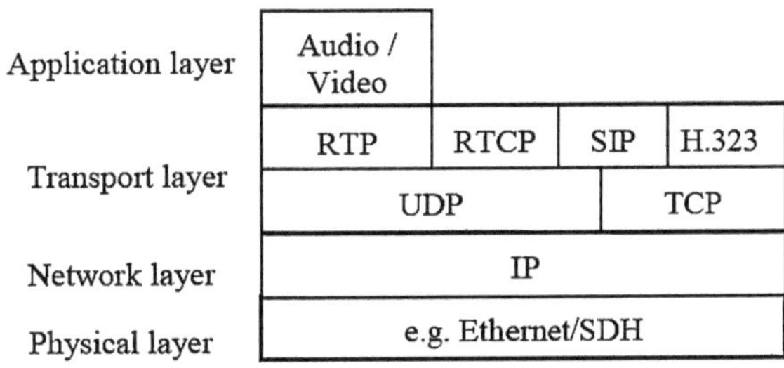

Fig. 2.3 VoIP protocol structure

	Audio / Video			
Application layer				
	RTP	RTCP	SIP	H.323
Transport layer	UDP			TCP
Network layer	IP			
Physical layer	e.g. Ethernet/SDH			

master–slave control-signaling protocols such as Media Gateway Control Protocol (MGCP) [26-27], and Megaco/H.248 [28].

2.2 Quality of Service

Quality of Service (QoS) is a measure of the voice quality experienced by the user. The network service provider uses it for bandwidth management over the IP network in order to ensure that transmission resources consistent with the expected QoS are available. Management of network resources is becoming increasingly important as more services are added on to the Internet.

VoIP becomes an attractive and common solution for the future, since the packet-switching technology has several advantages in both cost and architectural aspects over the circuit-switching technology. However, questions remain as to whether the voice quality provided in VoIP networks can meet the high standards provided by the PSTN that users have become accustomed to, and would expect from any competing service. The quality of speech perceived by the VoIP user is ultimately determined by parameters such as delay, jitter and packet loss [29].

2.2.1 A. Delay

Due to the interactive nature of voice communication, delay becomes a primary parameter of concern in the QoS measure for VoIP networks. It is composed of transmission delay, queuing delay, processing delay and propagation delay [30]. The transmission delay is dependent on the channel capacity in bits per second (bps). Queuing delay is the time the packets are queued in the buffer before being processed. Processing delay is incurred at the end points, e.g., in processing packet headers, and in coding/decoding voice signals. Propagation delay depends on the distance traveled and the transmission medium, such as coax, fiber, or wireless channel. The propagation delay is generally negligible when compared to the other components of delay in an end-to-end VoIP scenario.

International Telecommunications Union-Telephony (ITU-T) Recommendation G.114 [31] provides one-way transmission delay specifications for voice. The specification is presented in Table 2.1. It has been shown that a mouth-to-ear delay of over 150 milliseconds (ms) is intolerable to VoIP users, and the delay between successive packets must be lower than 20 ms for uninterrupted and smooth hearing [33]. Studies have shown that several techniques such as Weighted Fair Queuing, Weighted Round Robin, Priority Round Robin, Priority Queuing, or Class-based Queuing [34], can reduce the network delay.

In this book, we largely focus on the impact of queuing delay on VoIP networks. Since voice traffic has higher priority over data traffic, the queuing behavior of the voice packets is analyzed independently from the data packets. It is

Table 2.1 Delay specifications for voice [31]

Delay	Impact	Pre-Condition
Below150 ms	Acceptable for most user applications	Adequate echo control for connections of one-way delay more than 25 ms, as described in G.131 [32]
150–400 ms	Acceptable for international calls	
Above 400 ms	Unacceptable for general network planning purposes, especially in the case of transporting voice in packet switched networks.	

well known that an aggregate of voice (and Constant Bit Rate) video sources is reasonably accurately modeled by a Poisson arrival process and that queuing delays in consecutive nodes are more or less statistically independent [35]. Accordingly, we model two scenarios represented by the M/M/1 and M/D/1 queuing disciplines, and develop one method of calculating the throughput under a specified threshold of the total queuing delay through a VoIP network of N nodes. In addition, the analytical results addressed are used in scaling resources in a VoIP network for different thresholds of acceptable delays.

2.2.2 B. Jitter

Jitter is delay variation. It can lead to the gaps in the play out of the voice stream. The jitter can be compensated by maintaining a play out buffer on the receiver side [36], which processes the incoming packets in such a way that packets arriving earlier than average are buffered for a longer period than those arriving later. This means that the received voice stream can be recovered at a steady rate. In addition, arriving voice packets that exceed the maximum length of the jitter buffer are discarded.

2.2.3 C. Packet Loss

From an end-to-end point of view, the overall packet loss includes the network packet loss and the packet loss due to late arriving packets that are dropped at the jitter buffer. Packet loss can introduce audio distortion because of voice skips and clipping. Moreover, it can also introduce considerable impairment to voice signals. Typically, a packet loss rate of more than 5% is unacceptable for the VoIP users [37]. In order to reach the equivalent level of voice quality in a PSTN, the threshold rate of packet loss should be set below 1% in VoIP networks.

There are two methods to correct packet loss in packet switched networks. One is to use Forward Error Correction (FEC). The other is to use the packet loss concealment (PLC) algorithm [38]. The FEC method requires data redundancy and allows the reconstruction of lost data [39, 40]. The disadvantage of this approach is that it causes overhead bits and, therefore, additional delay. The PLC method, as implied in its name, conceals the packet loss. It uses a variety of techniques to recover the missing packets, such as silence substitution, packet repetition, waveform substitution, and pitch waveform replication [41].

2.3 VoIP Implementation

Network impairments affect the voice quality [42]. This section describes a set up in the laboratory that measures the voice quality under different kinds of network impairments.

2.3.1 VoIP Test Bed

A SIP-based VoIP test-bed is implemented as shown by interconnecting the University of Oklahoma-Tulsa (OU-Tulsa) to sip.edu by a peering arrangement. CISCO 2600 routers are configured as media gateways, and MySQL 4.0.21 open source database as the location database. SIP Express Router (SER) from http://www.iptel.org is installed and configured as the SIP proxy server. Figure 2.4 shows the implemented VoIP infrastructure for the OU-Tulsa TCOM Lab.

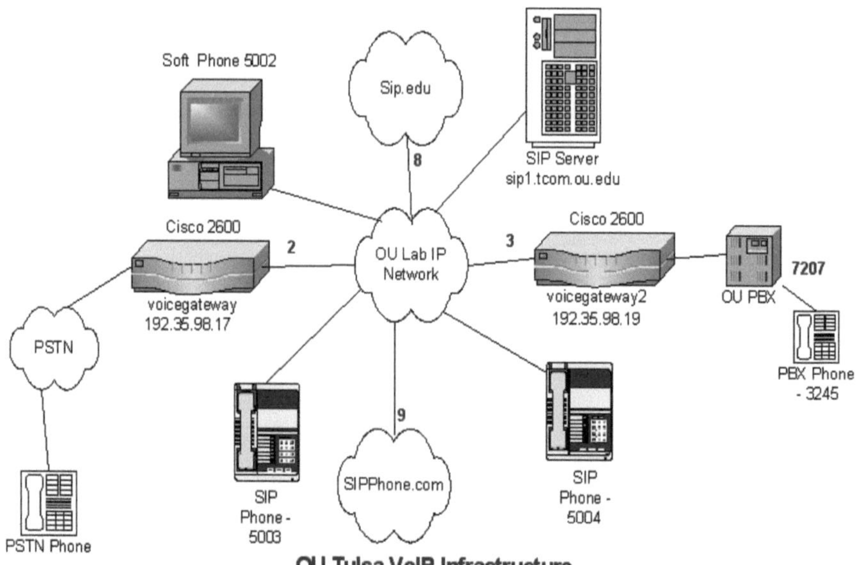

Fig. 2.4 OU TCOM-Lab VoIP infrastructure

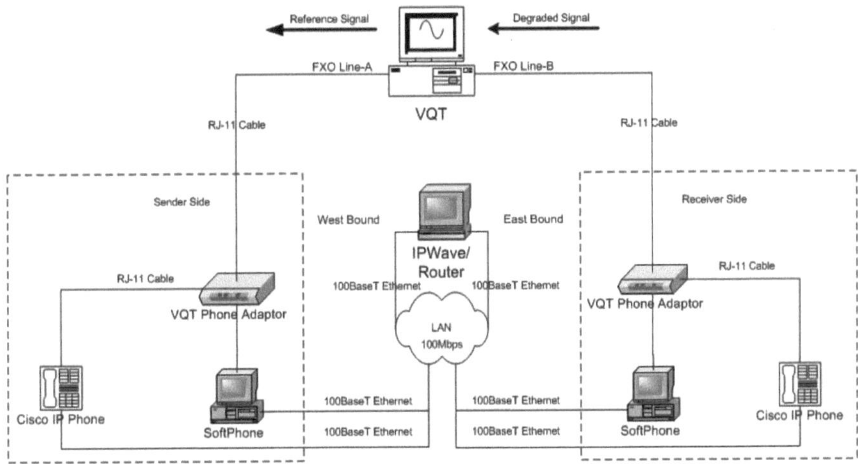

Fig. 2.5 Physical setup of the test-bed

In order to investigate the effects of various network impairments on the voice channel in the VoIP networks, the following test-bed to measure the perceived speech quality can be used. The test-bed consists of IPWave [43], Voice Quality Tester (VQT) [44] and the original VoIP network, as shown in Fig. 2.5.

IPWave and Agilent VQT are running on the Windows-NT operating system. IPWave is a network impairment generator to emulate the real world network conditions. It divides the network into Westbound and Eastbound and functions as a gateway. It introduces various network impairment conditions to the IP traffic from Westbound to Eastbound and vice versa. These impairments include packet loss, delay, jitter, out-of-order packets, and error in packets. The Agilent VQT is an objective speech quality measurement system used to predict the MOS of the perceived speech quality by means of the Perceptual Speech Quality Measurement (PSQM) algorithm [45]. In order to connect the FXO line of the Agilent VQT to either hard phone headset or soft phone PC's sound card, the Agilent VQT phone adapter [46] is used.

2.3.2 Measurement of Voice Quality

Voice quality is inherently subjective because it is determined by the listener's perception. The subjective voice quality is measured by objective measurement techniques, using the Mean Opinion Score (MOS) parameter.

The perceived speech quality is measured in the way shown in Fig. 2.6. The Agilent VQT captures the perceptual domain representation of two signals, namely, a reference signal that is input to the test-bed, and a degraded signal that is the output of the test-bed. It uses Perceptual Speech Quality Measurement (PSQM)

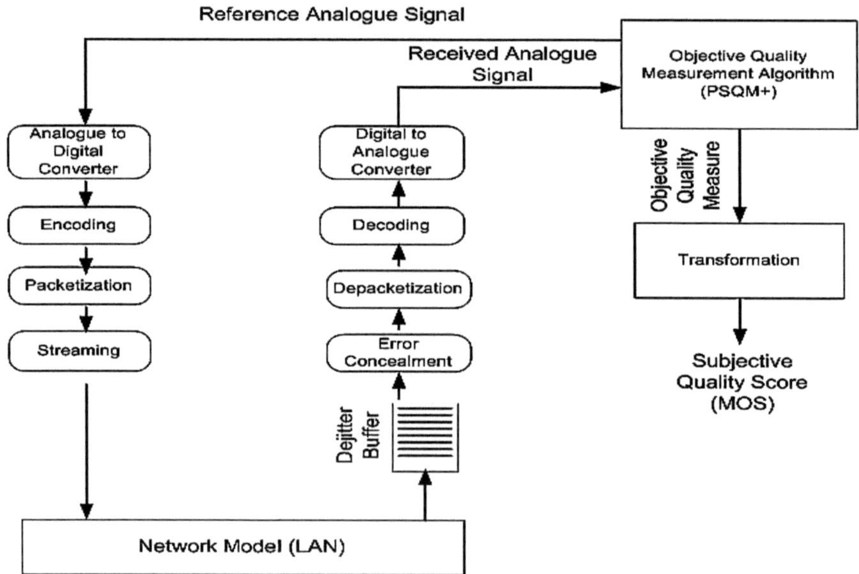

Fig. 2.6 Block diagram of the measurement [47, 48]

to analyze the voice quality in terms of MOS, which is widely accepted as a norm for voice quality rating.

Different experiments were conducted to determine the influence of packet loss on the voice quality. In this measure, we only apply one codec (G.711-μLaw) selected from the Cisco hard phone. We observe the impact on quality of a number of factors: periodic packet loss, random packet loss and burst packet loss. The results of these three loss models are shown in Figs. 2.7, 2.8 and 2.9, respectively. By comparing the three figures, we can see that the voice quality decreases as the amount of packet loss increases. It also shows that burst packet loss has more influence on the perceived voice quality.

2.4 Session Initiation Protocol

This part introduces the Session Initiation Protocol (SIP) specification and provides important aspects of SIP application in Voice over IP networks.

2.4.1 Background

The Session Initiation Protocol owes its origin in 1996 to the Internet Engineering Task Force (IETF) in order to distribute multimedia content. Since SIP was standardized to be adopted for Voice over Internet Protocol (VoIP) in 1999 as

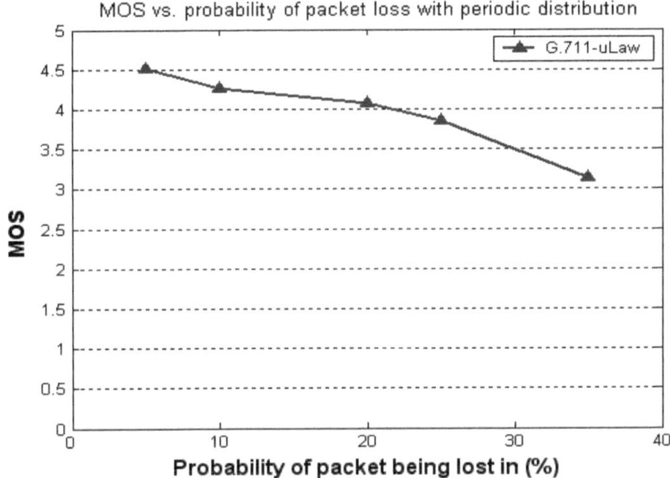

Fig. 2.7 Periodic loss model

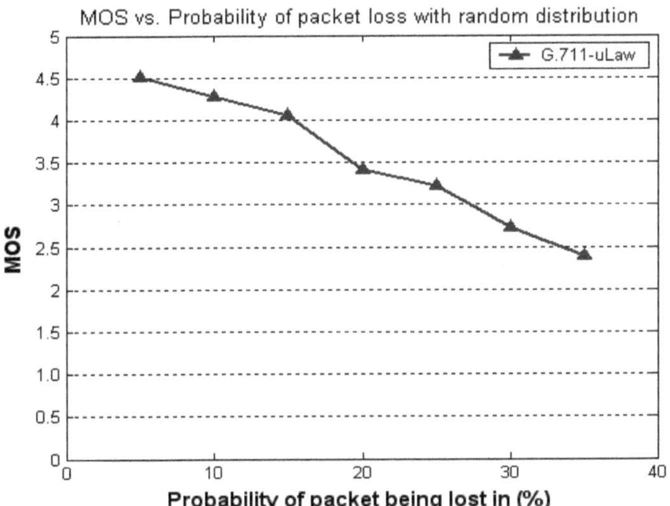

Fig. 2.8 Random loss model

RFC2543, it has evolved significantly and now it covers a wide range of real-time collaboration functionalities [49]. In this chapter, we will only focus on the latest standard RFC3261.

SIP is an end-to-end, client-server session signaling protocol. It is designed to establish presence, locate users, set up, modify and tear down voice and video sessions across the packet-switched networks. Borrowing from the ubiquitous Internet protocol, such as the hypertext transfer protocol (HTTP) and simple mail transfer protocol (SMTP), SIP is text-encoded, programmable, and highly

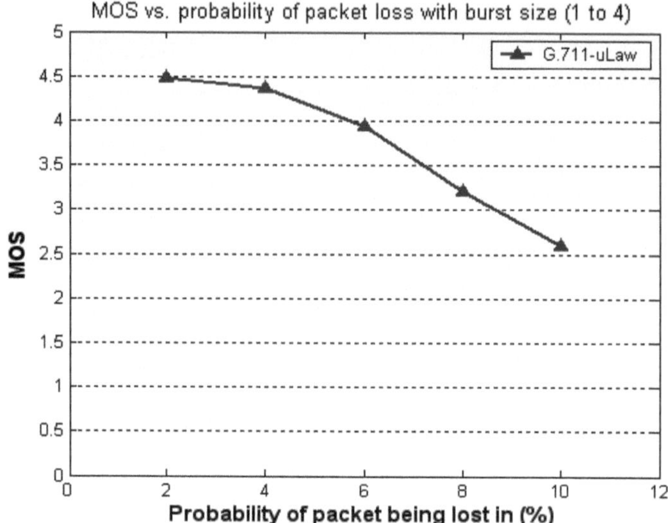

Fig. 2.9 Burst loss model

Fig. 2.10 A Cisco 7960 SIP
Phone

extensible [50]. Due to its simplicity and extensibility, as well as the newly created features, SIP is not limited to IP telephony. SIP messages can convey arbitrary level of signaling payload, session description, instant messages, JPEGs, and any MIME type. SIP uses Session Description Protocol (SDP) [51] for media description.

2.4.2 SIP Network Elements

2.4.2.1 A. User Agent

User agents are end entities in SIP-based Networks to connect each other and negotiate session parameters. User agents can be both hardware and software. For example, in the SIP-based VoIP testbed in the lab, we used a Cisco 7960 SIP phone as shown in Fig. 2.10. It usually, but not necessarily, resides on a user's computer in form of a user application [52]. It can also be a PSTN gateway, a cellular phone, a PDA and so on.

In terms of functionalities, a UA can be categorized into User Agent Client (UAC) or User Agent Server (UAS). UAC and UAS are logically separated but physically combined in the same end point. UAC works on behalf of the client to originate the call and receive the response, whereas the UAS functions on the behalf of the server to listen to the incoming calls and to respond to request. For example, in order to initiate a call session, an INVITE message is sent by the caller's UAC, and received by the callee's UAS. On the other hand, in order to terminate the session, a BYE message is sent by callee's UAC and received by the caller's UAS.

2.4.2.2 B. SIP Server

Based on the functionalities, SIP servers are logically classified into three components as Registrar, Proxy Server, and Redirect Server.

Registrar is one of the SIP servers used to initialize and keep record of the user agent. It accepts the REGISTER requests and maintains the information of the users' AoR (Address of Record) including various kinds of SIP URL addresses binding to the same user. Registrar also indicates the current address as the first priority where the user wants to send the request and receive the response [53].

Proxy server plays a very important role in processing the SIP signaling messages. It receives the request from the users and looks up in the location server, where all the records of the users are kept, to find the destination address. And then the SIP server forwards the request by interpreting, and modifying certain parts of the INVITE message, such as Via. Proxy servers can be classified as stateful proxy server or stateless proxy servers.

2.4.3 SIP Messages

SIP messages are divided into two types depending on the direction of the messages. The SIP message sent from the client to the server is the Request message,

Table 2.2 Request methods example	Method	Description
	INVITE	Initiates a call, changes call parameters (re-INVITE)
	ACK	Confirms a final response for INVITE
	BYE	Terminates a call
	CANCEL	Cancels searches and "ringing"
	OPTIONS	Queries the capabilities of the other side
	REGISTER	Registers with the Location Service
	INFO	Sends mid-session information that does not modify the session state

Table 2.3 Response example

Type	Class	Description	Examples	
			Code	Meaning
Provisional	1xx	In Progress	100	Trying
			180	Ringing
Final	2xx	Success	200	OK
	3xx	Redirection	300	Multiple choices
			301	Moved permanently
			302	Moved temporarily
	4xx	Client Error	400	Bad request
			401	Unauthorized
			403	Forbidden
			408	Request time-out
			480	Temporarily unavailable
			481	Call/Transaction does not exist
			482	Loop detected
	5xx	Server Error	500	Server error
	6xx	Global Failure	600	Busy everywhere
			603	Decline
			604	Does not exist anywhere
			606	Not acceptable

while that from the server to the client is the Response message. Tables 2.2 and 2.3 give examples of the Request and Response SIP messages, respectively.

SIP messages consist of three main parts: start line, header, and message body. Each SIP message begins with a start line to convey the message type and the protocol version. SIP headers are borrowed from the syntax and semantics of HTTP header fields, to convey more message attributes. The message body can use either Session Description Protocol (SDP) or Multipurpose Internet Mail Extensions (MIME). Here is an example of the INVITE message:

```
INVITE sip:bob@nice.com SIP/3.0
   Via: SIP/3.0/UDP 192.2.4.4:5060
   To: Bob < sip:555-6666@nice.com>
   From: Aline < sip:555-1234@nice.com > ;
   tag = 203 941 885
   Call-ID: b95c5d87f7721@192.2.4.4
   Cseq: 26 563 897 INVITE
   Contact: < sip:555-1234@192.2.4.4>
   Content-Type: application/sdp
   Contact-Length: 142

   v = 0
   o = Alice 53655765 2353687637 IN IP4
   128.3.4.5
   s = Call from Alice
   c = IN IP4 192.2.4.4
   M = audio 3456 RTP/AVP 0 3 4 5
```

2.4.4 SIP Transactions

A SIP transaction is a sequence of SIP messages ranging from the request to all responses to that request. SIP is transactional, because the SIP messages are arranged into transactions, although they are sent independently. SIP transactions have both client and server sides. In each side, there are two types known as an INVITE transaction, where the request is an INVITE, and the non-INVITE transaction. Unlike the INVITE transaction, a non-INVITE transaction only has a single 2xx response, without ACK or other special handling. Figure 2.11 gives examples of two SIP transactions. In Trans #1, the ACK is not considered part of the transaction since the response was a 2xx. While in Trans #2, the ACK is included in the transaction only if the final response is not a 2xx response.

As addressed in RFC3261 [25], the transaction identifier is expressed as the branch parameter inside the Via header fields. However, since the previous SIP RFC2543 calculates the transaction identifier as the hash of all important message header fields (that included To, From, Request-URI and CSeq) [54], a compatible feature should be provided for backward support.

2.4.5 SIP Dialogues

SIP dialog represents a peer-to-peer relationship between two end user agents. Also shown in Fig. 2.11, the two transactions are not treated independently, but related in such a way that they are identified as belonging to the same *dialog*.

Fig. 2.11 SIP transactions and dialogs

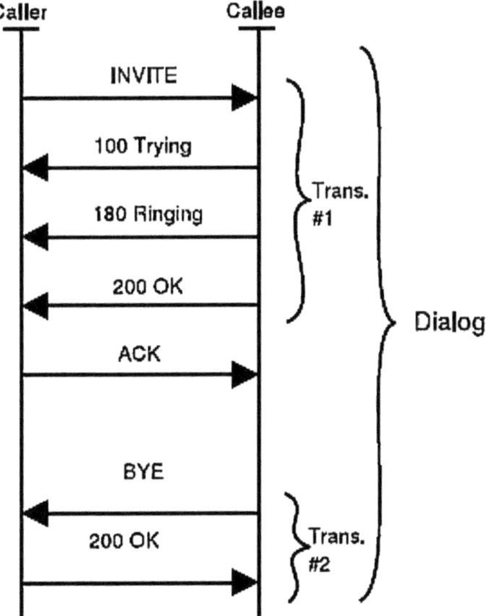

Being identified by From and To tags and the Call-ID, SIP Dialogs facilitate proper sequencing and routing of messages between the user agents [25]. Also, the command sequence (*Cseq*) contains an integer and a method name. This *Cseq* number is incremented for each new request, which actually means that the *CSeq* number identifies a transaction. To some degree, *a dialog is a sequence of transactions* [52].

2.4.6 Typical SIP Scenarios

To understand SIP signaling, two scenarios to illustrate the SIP message flow are presented in the following.

One is a redirection scenario as shown in Fig. 2.12. Upon receiving the INVITE message from the user agent A, the redirect server responds with 302 (Moved Temporarily), indicating the user agent B is temporarily available at an alternate address expressed in the Contact header. Sometimes, the duration of validity of these addresses is also included. After returning the acknowledgement to the redirect server, the user agent A sends a second INVITE message directly to the user agent B, by using the routing information pushed back from the redirect server. With the aid in locating the target of the request from the redirect server, the procedure becomes simple and quick. In other words, the redirect server results in a high level of performance. It is worth noting that the second INVITE message has a different CSeq value from the first INVITE message; however, the To and From headers, Call-ID, and dialog identifiers remain the same. The following sequence of signaling is common in each scenario: once the user agent B picks up

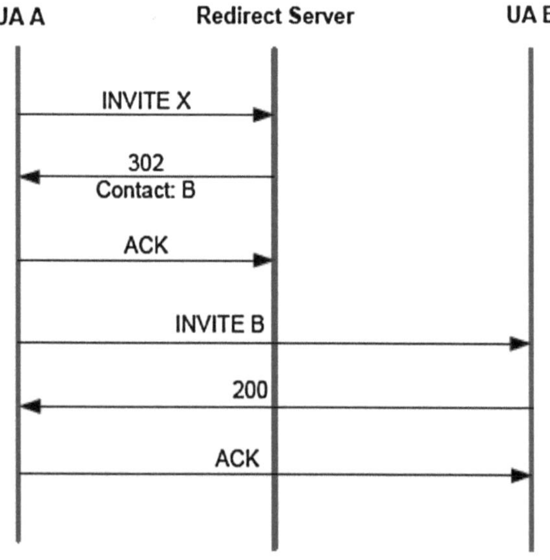

Fig. 2.12 Signaling flow with redirect server [55]

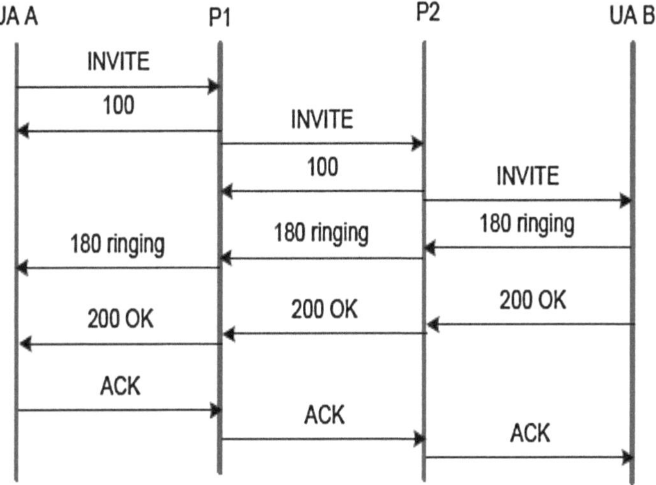

Fig. 2.13 Signaling flow with proxy server [55]

the phone, the 200 OK message is sent back to the user agent A, and the media flow is established after the user agent B receives the acknowledgement.

The other scenario, as shown in Fig. 2.13, is that the request traverses multiple proxy servers before reaching the destination. The main difference from the first scenario is that after making the routing decision, each intermediary proxy server modifies the INVITE message and then forwards it to the next proxy server. The response routes through the same set of proxies in the reverse order.

2.5 Summary

This chapter has provided a brief overview of VoIP networks from different perspectives. Laboratory implementation and studies on the measurement of voice quality have been discussed. The popular VoIP signaling procedure SIP has been described in detail. In the next section, we will present the analytical model for delay-throughput analysis adopted in this book.

Chapter 3
Traffic Characterization

Abstract This chapter presents the characterization of traffic in packet-switched networks. It provides an overview of queuing theory, including the terminology and notations, which will be used in the analyses presented in the succeeding chapters. In order to study the relationship between delay and throughput in the context of VoIP networks, applicable queuing theory models are developed and presented. Analyses of these models are discussed in Chaps. 4 and 5.

In order to analyze a telecommunication network, that carries both data and voice traffic, an analytical model is needed to capture the main characteristics of the system. Since the behavior of a system is characterized in conjunction with the traffic it handles, modeling of the traffic it handles is necessary and important because it can help find the optimal network configuration and sizing without having to build prototypes [56]. Mathematical tools commonly applied in the study of stochastic processes, queuing theory, and simulation are used to model the traffic. Since this book is largely focused on understanding the queuing delay in VoIP networks, results from queuing theory are mainly used to analyze the VoIP network.

Despite the fact that data networks (such as the Internet) are drastically different from the legacy public switched telephone networks, the long held paradigm in the communication and networking research community has been that data traffic— analogous to voice traffic—is adequately described by certain Markovian models which are amenable to analysis and efficient control. Newer models of characterizing data traffic have emerged as in [57]. However, conventional models of analysis are potentially sufficient to draw the conclusions we wish to develop in the succeeding chapters. This book, accordingly, models Voice over IP networks using conventional techniques in a way that relates a predefined quality of service consistent with optimal utilization of network resources. Furthermore, we propose a new parameter for capturing the quality of service.

P. K. Verma and L. Wang, *Voice over IP Networks*,
Lecture Notes in Electrical Engineering, 71, DOI: 10.1007/978-3-642-14330-4_3,
© Springer-Verlag Berlin Heidelberg 2011

3.1 Packet-switched Network Model

The traditional telephone network has been designed as a hierarchical system constituted by the local exchange (LEX) and transit exchanges (TEX). LEX is the subscriber switch which connects the individual to the network. TEX is a transit exchange connecting to several LEXs. The TEX themselves form a hierarchy. The highest-level switches of this hierarchy are fully interconnected with each other.

A packet-switched data network is a packet distribution network following the store-and-forward principle at the packet level [58], as shown in the Fig. 3.1. It can be modeled as a stochastic flow system, where the flow is the packet and the channel is the network link. The data to be sent is segmented into packets and can be transmitted through different paths through the network. The packet-switched network introduces queuing delay at each node. Voice packets are much more sensitive to delay and jitter than data packets. Our aim in this book is to reduce the queuing delay by analyzing performance through the application of queuing theory and appropriate provisioning of network resources.

3.2 Queuing Model

In this section, a mathematical model is described to represent packet-switched data networks that support voice traffic [59]. An upper bound for the end-to-end delay for a specified percentage of packets is chosen as an objective measure of the quality of service. This forms the basis of our analysis. It is a better measure of performance than the average delay commonly used in analysis. As mentioned in Chap. 2, there are many sources of the packet delay. As mentioned earlier, the other delays are mostly uncontrollable delays such as the propagation and

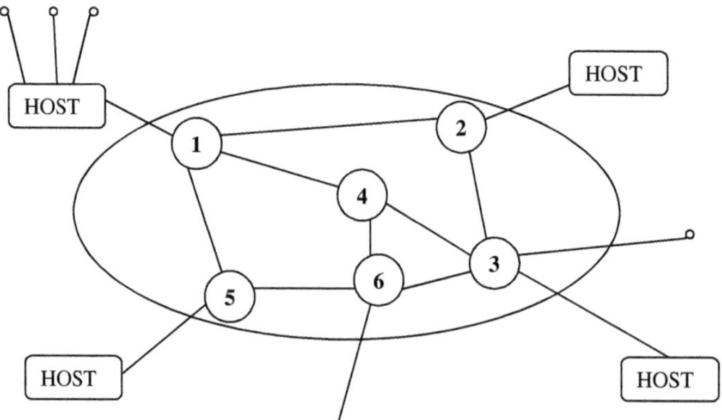

Fig. 3.1 Packet-switched data network

processing delays. Queuing delay is the variable component of the end-to-end delay causing considerable degradation in real-time voice communication.

3.2.1 Queuing Specification

The queuing model used in this book was developed by D.G. Kendall [60] and represented as A/B/n, a notation that is widely used in literature. A represents the distribution of the arrival process, B implies the distribution of the service time, and n is the number of servers. It needs to be mentioned that B is directly related to the length of the packets. To provide more information on the queuing system, a more complete specification is formatted as A/B/n/K/S/X, where K is the buffer size or the capacity of the link, S is the population of the arrival packets, and X denotes the queuing discipline.

The standard notations for frequently used traffic processes are shown in the Table 3.1.

3.2.2 Assumptions of the Queuing Model

Analytical modeling of real systems for queuing analysis is a complicated task. To simplify the analysis and yet get a realistic evaluation of the end-to-end delay characteristics of packetized voice traffic [61], the following assumptions have been made:

- All channels are assumed to be noiseless [62]. The data rate of transmission over the channel does not exceed the channel capacity.
- The buffer size is unlimited. It implies that the storage capacity at each node is infinite to provide waiting positions in the queues and sustain transient congestions.
- The transmission is considered only over one or more point-to-point links. Each packet is delivered to its single destination.
- The processes of the arrival and the service time distribution are mutually independent. The assumption of preservation of the Poisson arrival process at an intermediate node is a strong assumption made throughout this book. We justify

Table 3.1 Kendall's notations

Notation	Description
M (Markovian)	Exponential time intervals (Poisson arrival process, exponentially distributed service times)
D (deterministic)	Constant time intervals
Ek	Erlang-k distributed time intervals (E1 = M)
G (general)	Arbitrary distribution of time intervals (may include correlation)

it based on the following observation. At an intermediate node, while the message arrival process from a single node might not be Poisson, the collective inter-arrival times and the lengths of messages generated by the entire population of subscribers arriving from different adjoining nodes exhibit independence. This is because the length of a message generated by any particular subscriber is completely independent of the arrival times of messages generated by the other subscribers [62]. It is actually the multiplicity of paths that lead to a single node and multiplicity of exits off that node that considerably reduces the dependency between inter-arrival times and lengths of messages as they enter and exit the various nodes within the network. Extensive simulations of networks in [62] validate this simplifying assumption that makes the mathematical analysis tractable.

- The arrival process of the packet is Poissonian, and the service discipline is characterized by a negative exponential distribution.
- For multiple hop traffic, the waiting times at each node are independent. With this assumption (approximation), the distribution of the end-to-end delay can be derived by the convolution of the waiting times at each node [63].

3.2.3 Statistical Properties of Traffic

As stated in the last section, the traffic arriving at a node is assumed to follow the Poisson discipline. The Poisson process is generally considered to be a good model for the aggregate traffic from a large number of similar and independent users [64]. Although there is no quantitative data to determine the actual distribution of the real traffic in all cases, certain data obtained by Molina [65] for telephone traffic correspond very well to the assumption [62]. Table 3.2 shows characteristics of distributions associated with the Poisson arrival process and exponential distribution of service time.

Regarding the service time distribution, two cases are considered. The first case is the M/M/1 model, where the service time follows the exponential distribution with the parameter μ. It also indicates that the packet length is exponentially distributed. The M/M/1 model has long been the typical model for queuing analysis

Table 3.2 Commonly used distributions

Distribution	Exponential	Erlang-k	Poisson
Description	Interval between two events or from a random point until the next event	Time interval until the k'th event	Number of events in a time interval t
Formula	$f(t) = \lambda e^{-\lambda t},\ t \geq 0$ $m_1 = \frac{1}{\lambda}$ $\sigma^2 = \frac{1}{\lambda^2}$	$f(t\|k) = \frac{(\lambda t)^{k-1}}{(k-1)!}\lambda e^{-\lambda t}$ $t \geq 0$ $m_1 = \frac{k}{\lambda},\ \sigma^2 = \frac{k}{\lambda^2}$	$f(x\|t) = \frac{(\lambda t)^x}{x!}e^{-\lambda t}\ t \geq 0$ $m_1 = \lambda t$ $\sigma^2 = \lambda t$

due to its simplicity, for such widely varying applications as the terminal to computer communications, shared Local Area Networks, and airline reservations [66]. However, we also place special emphasis on the second case that is the M/D/1 model, with constant service time. The M/D/1 model is important since streaming traffic such as that associated with human conversation might be packetized as successive packets of constant length. Additionally, the ATM technology uses packets of fixed length.

3.3 Analysis of the Delay Bound

Most of the work presented in this book is devoted to improving the throughput within a predefined upper delay bound, in particular, the queuing delay. The waiting time distribution influences the upper delay bound. For simplicity, we assume the queue to follow the discipline First-In First-Out (FIFO), also called First-Come First-Served (FCFS). Packets arriving first to the node will be served first. Also it is assumed that there is single server at each node. In the following, the waiting time distributions for both M/M/1 and M/D/1 models are derived respectively. Table 3.3 summarizes the notations used in queuing analysis and delay calculations throughout the book.

3.3.1 The M/M/1 Model

Let us consider a typical case. Assume the number of packets waiting in the queue at the arrival time is denoted by X^*. In other words, it is the queue length as seen by an arriving packet. The Poisson arrival process has the PASTA-property (Poisson Arrivals See Time Averages),

$$P\{X^* = i\} = P\{X = i\} = \pi_i, \tag{3.1}$$

where X is the number of packets in the node at an arbitrary time.

Due to the memoryless property of the exponential distribution, the service time S_1^* of the packet in service also follows exponential distribution with the parameter μ, and is independent of other packets in the queue. Service times of the packets waiting in the queue are identically and independently distributed. For the assumed the FIFO queuing discipline, the total waiting time for the typical packet is given as,

$$W = S_1^* + S_2 + \cdots + S_i, \tag{3.2}$$

where $\tau_1 = S_1^*$ and $\tau_n = S_1^* + S_2 + \cdots + S_i$, $n \geq 2$.

It is known that the Poisson distribution is a point process [67].

Table 3.3 Notations used

Symbol	Meaning
n	Number of the nodes in the network
h	Number of the hops in the network
λ	Mean packet arrival rate
$\frac{1}{\mu}$	Mean length of each packet (bits)
C	Transmission capacity of each link (bps)
μC	Service rate (packets per second)
ρ	Utilization factor $\frac{\lambda}{\mu C}$
W_n	End-to-end queuing delay (n hops)
t	End-to-end threshold delay
$F_W(t)$	Distribution of waiting time $P\{W \le t\}$
$f_W(t)$	Probability density function
γ_n	Throughput of n-hop network
D	End-to-end delay
D_i	Delay introduced by the ith hop
$D^{(i)}$	The ith moment of the delay
σ_D^2	Jitter (delay variance)

It can be shown that,

$$P\{W = 0\} = P\{X^* = 0\} = \pi_0 = 1 - \rho. \tag{3.3}$$

Since $X^* = i$, we have $W > t \Leftrightarrow \tau_i > t$. Hence,

$$P\{W > t\} = \sum_{i=1}^{\infty} P\{W > t | X^* = i\} P\{X^* = i\} = \sum_{i=1}^{\infty} P\{\tau_i > t\} \pi_i = \sum_{i=1}^{\infty} P\{\tau_i > t\}(1 - \rho)\rho^i.$$

$$\tag{3.4}$$

The Poisson process can be also seen as a counter process $A(t)$ corresponding to τ_n, from which it follows that,

$$\tau_i > t \Leftrightarrow A(t) \le i - 1. \tag{3.5}$$

On the other hand, it is already known that $A(t)$ follows the Poisson distribution with parameter μt. From Eqs. 3.4 and 3.5, we have,

$$\begin{aligned}
P\{W > t\} &= \sum_{i=1}^{\infty} P\{\tau_i > t\}(1 - \rho)\rho^i = \sum_{i=1}^{\infty} \sum_{j=0}^{i-1} \frac{(\mu t)^j}{j!} e^{-\mu t}(1 - \rho)\rho^i \\
&= \rho \sum_{j=0}^{\infty} \frac{(\mu t \rho)^j}{j!} e^{-\mu t}(1 - \rho) \sum_{i=j+1}^{\infty} \rho^{i-(j+1)} \tag{3.6} \\
&= \rho \sum_{j=0}^{\infty} \frac{(\mu t \rho)^j}{j!} e^{-\mu t} = \rho e^{\mu t \rho} e^{-\mu t} = \rho e^{-\mu(1-\rho)t}.
\end{aligned}$$

Equation 3.6 gives the probability that the waiting time is greater than a stipulated value t. It is interesting to find that the distribution of the waiting time W can be presented as the product of two independent random variables; one follows the Bernoulli distribution with parameter ρ and the other follows the exponential distribution with parameter $\mu(1 - \rho)$. In other words,

$$W = JD, \quad \text{where } J \sim Bernoulli(\rho) \text{ and } D \sim \text{Exp}(\mu(1 - \rho)). \tag{3.7}$$

Thus we have the average waiting time as

$$E(W) = E(J)E(D) = \rho \frac{1}{\mu(1 - \rho)} = \frac{\rho}{\mu(1 - \rho)}. \tag{3.8}$$

After deriving the waiting time distribution, the throughput of the VoIP network in which all packets undergo a queuing delay not exceeding the threshold delay t, can be derived as:

$$\gamma = \lambda p, \tag{3.9}$$

where λ is the arrival rate of packets, p is the probability of the waiting time less than or equal to the threshold delay t, which equals $1 - P\{W > t\}$.

Also, p is referred to as the normalized throughput, i.e., throughput expressed as a function of the incident traffic, which is given as:

$$p = \frac{\gamma}{\lambda}. \tag{3.10}$$

This relationship described by (3.10) remains the same for the M/D/1 model.

3.3.2 The M/D/1 Model

Several formulas have been proposed to evaluate numerically the waiting time distribution for the M/D/1 model. For small waiting times, Eq. 3.11 given below is used for numerical evaluation. For larger waiting times, there are two different cases. With the integral values of the waiting time t, Eqs. 3.12 and 3.13 are used for calculation. With non-integral values of waiting time, the waiting time distribution is expressed in terms of integral waiting times, as shown in Eq. 3.14.

As given by Erlang in 1909, the distribution function of waiting time can be written in a closed form [68],

$$P(W \le t) = (1 - \lambda) \sum_{j=0}^{T} \frac{[\lambda(j - t)]^j}{j!} e^{-\lambda(j-t)}, \tag{3.11}$$

where $t = T + \tau$, i.e., $T = \lfloor t \rfloor$, where $\lfloor t \rfloor$ is the largest integer less than or equal to t.

Iverson [69] has shown that for an integral value of t, we have,

$$P\{W \leq t\} = p(0) + p(1) + \cdots + p(t), \tag{3.12}$$

where $p(i)$ is the state probability. To be accurate, the state probabilities are calculated by using a recursive formula based on Fry's equations of state as [70],

$$p(i+i) = \frac{1}{p(0,h)}\left\{p(i) - [p(0) + p(1)]p(i,h) - \sum_{j=2}^{i} p(j) \cdot p(i-j+1,h)\right\}. \tag{3.13}$$

For the non-integral value of t, it is expressed as $t = T + \tau$. Therefore, we have,

$$P(W \leq T + \tau) = e^{\lambda\tau} \sum_{j=0}^{T} \frac{(-\lambda\tau)^j}{j!} P\{W \leq T - j\}. \tag{3.14}$$

Note that $P\{W \leq T - j\}$ can be calculated using Eq. 3.11.
In this book, Eq. 3.11 is applied for analysis of the M/D/1 system.

3.3.3 Comparison of M/M/1 and M/D/1 Models

Figure 3.2 shows the distribution of the queuing delay for both single M/M/1 and M/D/1 models. The parameter is different values of λ. Without loss of generality,

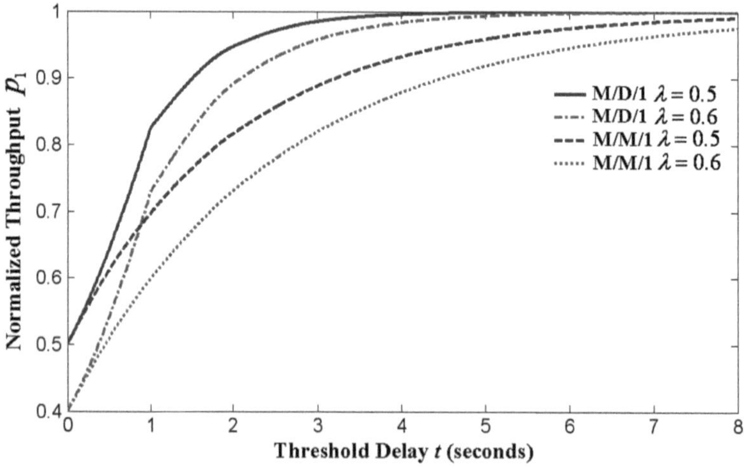

Fig. 3.2 The normalized throughput of both M/M/1 and M/D/1

the mean service time is scaled to the unit 1. As shown in the Fig. 3.2, the similarities of both models are:

- The normalized throughput increases with an increase in the threshold delay, reaching 100% asymptotically.
- Both M/M/1 and M/D/1 systems consistently show a higher normalized throughput for a smaller λ or, equivalently, smaller ρ.

The difference between M/M/1 and M/D/1 is that for a given value of λ, the M/D/1 system has higher normalized throughput than the M/M/1 system.

3.4 Summary

This chapter has provided a brief overview of traffic modeling using conventional queuing theory that is needed in the analyses that follow. The waiting time distributions for both M/M/1 and M/D/1 disciplines have been discussed. In the following chapter, a VoIP network modeled as an M/M/1 system is presented. A closed form solution relating the impact of the bounded delay on the throughput is developed.

Chapter 4
Impact of Bounded Delays on Resource Consumption in Packet-Switched Networks with M/M/1 Traffic

Abstract This chapter presents a closed form solution relating the impact of bounded delays on throughput in VoIP networks modeled by M/M/1. Traffic that exceeds the delay threshold is treated as lost throughput. The results addressed can be used in scaling resources in a VoIP network for different thresholds of acceptable delays. Both single and multiple switching points are addressed. Simulations support the analytical findings. The contents of this chapter have been published, in part, in [71].

4.1 Introduction

Unlike the PSTN, the Internet was not specifically designed for voice communication. As such, VoIP calls can suffer from delay, jitter or packet loss [72]. As described in Chap. 2 , the legacy circuit-switched network allocates dedicated bandwidth to each call resulting in virtually no delay due to queuing and no jitter. PSTN provides a consistently high level of voice quality, called toll quality [73]. VoIP packets in packet-switched networks undergo varying amounts of delay at each transit hop. Therefore, delay becomes a significant parameter in VoIP networks. The delay addressed in this chapter is the variable queuing delay [13], the time each voice packet has to wait at each router in the path of a VoIP connection.

The initial driver for VoIP was low cost. Because of the reasons cited above, performance considerations are an important aspect of VoIP's continuing penetration in the marketplace. There are, of course, other issues as well such as ease of deployment and maintenance. As mentioned before, the major issue addressed in this book is performance and consumption of bandwidth commensurate with predefined performance level.

P. K. Verma and L. Wang, *Voice over IP Networks*,
Lecture Notes in Electrical Engineering, 71, DOI: 10.1007/978-3-642-14330-4_4,
© Springer-Verlag Berlin Heidelberg 2011

4.1.1 Average Delay Versus Bounded Delay

The average delay of voice packets is an indicator of the QoS [74] of a VoIP network. However, the average delay is not meaningful in real-time human conversations. Suppose two services with the same average delay of 300 ms are available. One ranges from 100 to 500 ms, the other from 200 to 400 ms. The 200–400 ms-bounded delay system provides a much higher voice quality than the 100–500 ms-bounded delay system. A variation in delay causes jitter. In many ways, jitter is potentially the single largest contributor to degradation in the quality of voice. One way to control jitter is the use of a jitter buffer. A larger jitter buffer will result in the network's ability to compensate for larger delay. However, a large jitter buffer will increase the delay. This will affect the interactive quality of voice communication.

The upper delay bound is a key factor in the choice of the systems. Accordingly, from the Service Providers' point of view, a system that limits the upper bound of the queuing delay to a low value is very important. Reducing the upper bound of delay will reduce both the delay as well as the jitter. This is the approach adopted in this book.

In the mathematical model used in this book, we assume an infinite buffer at each node. Packets are thus not lost, but delayed. As mentioned earlier, we use delay to characterize the Quality of Service. While contemporary literature largely uses mean delay to characterize the QoS, we propose to use an upper bound of the delay as a measure of the same. Traffic that suffers a delay higher than the bound is considered lost and does not constitute effective throughput [75]. We treat the 'lost throughput' as being equivalent to a call that was blocked in a circuit-switched network [76].

4.1.2 Organization of this Chapter

This chapter addresses the impact of upper delay bounds on throughput in VoIP Networks. The rest of the chapter is organized as follows: Sect. 4.2 describes a simple IP telephony system based on the Session Initiation Protocol (SIP) and Sects. 4.3, 4.4 and 4.5, develop analytical results for the model we have used for VoIP networks. The models consist of one-hop, two hops or multiple hops. Section 4.6 presents the results of simulation that verifies the analytical results. Section 4.7 presents the conclusion of this chapter.

4.2 The SIP-based VoIP Network Model

Figure 4.1 shows an example VoIP network where the end points are connected to a LAN and VoIP calls are routed through the LAN, WAN or a gateway to the remote end point. SIP has been used as the signaling protocol in Fig. 4.1. Not

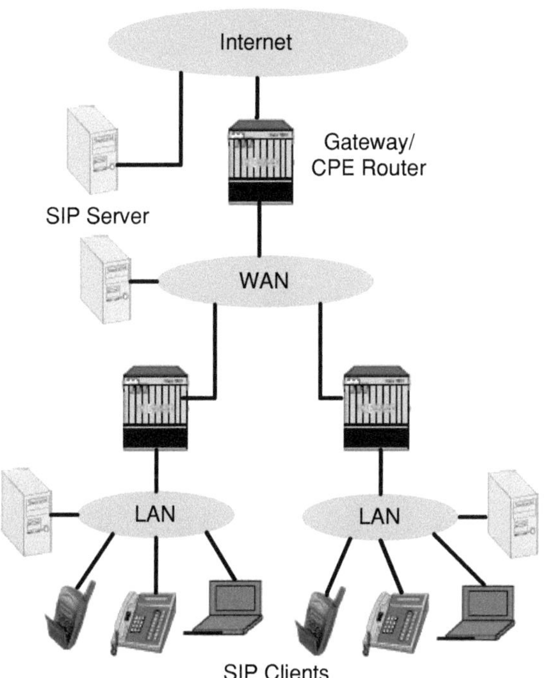

Fig. 4.1 SIP-based VoIP scenario

being limited to IP telephony, SIP messages can convey arbitrary signaling payload, session description, instant messages, JPEGs, any MIME types. SIP uses the Session Description Protocol (SDP) for media description.

4.3 A Single-Hop VoIP Network

As discussed in Sect. 4.2, a typical voice packet would pass through several hops before arriving at the destination. We consider the voice traffic served by a single hop as well as multiple hops.

Consider a LAN shown in Fig. 4.2 with a SIP server that functions as a VoIP network. We model the VoIP packets arriving at the SIP server as M/M/1 traffic [77, 78].

The distribution of the waiting time W can be written as [58]:

$$F_W(t) = P\{W \le t\} = 1 - \frac{\lambda}{\mu C}e^{-(\mu C - \lambda)t} = 1 - \rho e^{-\mu C(1-\rho)t} \tag{4.1}$$

The throughput of the single-hop switching system where all packets that undergo a queuing delay less than the threshold delay can be expressed as:

Fig. 4.2 Single-hop network

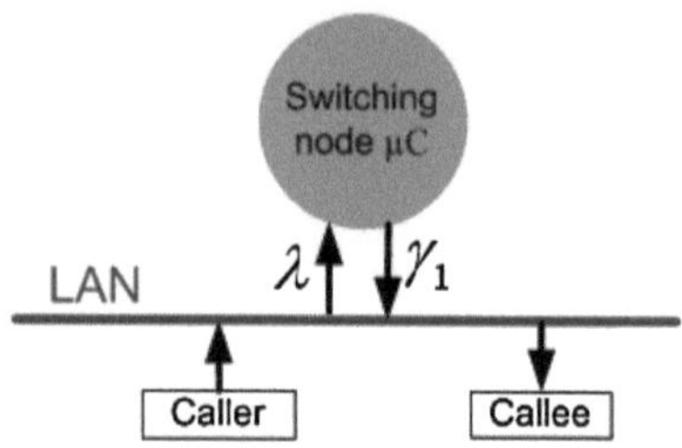

$$\gamma_1 = \lambda \cdot P\{W \le t\} = \lambda\left[1 - \frac{\lambda}{\mu C}e^{-(\mu C - \lambda)t}\right], 10 \tag{4.2}$$

4.3.1 Analysis of a Single-Hop VoIP Network

We first prove the following theorem that characterizes the behavior of a single-hop VoIP network.

Theorem 1 *The throughput γ_1 of a VoIP server, where the arriving traffic follows the M/M/1 discipline and all packets incurring a queuing delay higher than t are discarded, is maximized for a mean packet arrival rate λ_0 such that the following transcendental equation:*

$$\lambda_0(2 + \lambda_0 t) = \mu C e^{(\mu C - \lambda_0)t}$$

is satisfied. The maximized throughput under this condition is given by

$$\gamma_{\max} = \frac{\lambda_0(1 + \lambda_0 t)}{2 + \lambda_0 t}$$

Proof We note from Eq. 4.2 that γ_1 is continuous on the closed interval $[0, \mu C]$. Thus, if $\gamma_{\max}$ is an extreme value of γ_1 corresponding to λ_0 on that interval, then one of the following two statements is true: (a) $\gamma'(\lambda_0) = 0$, or (b) $\gamma'(\lambda_0)$ does not exist.

The first-order derivative of (4.2) is:

$$\frac{d\gamma_1}{d\lambda} = 1 - \frac{\lambda}{\mu C}(2 + \lambda t)e^{-(\mu C - \lambda)t} \tag{4.3}$$

which exists. The second-order derivative of (4.2) is:

$$\frac{d^2\gamma_1}{d\lambda^2} = -\left[\frac{2(1 + \lambda t)}{\mu C} + \frac{\lambda t(2 + \lambda t)}{\mu C}\right]e^{-(\mu C - \lambda)t} \tag{4.4}$$

which is negative. We can now obtain the maximum throughput $\gamma_{\max}$ by putting the first derivative of γ_1 equal to zero. In other words, γ_1 will be maximized for the specific λ_0 such that

$$1 - \frac{\lambda_0}{\mu C}(2 + \lambda_0 t)e^{-(\mu C - \lambda_0)t} = 0 \tag{4.5}$$

Equation (4.5) can be rewritten as,

$$\lambda_0(2 + \lambda_0 t) = \mu C e^{(\mu C - \lambda_0)t} \tag{4.6}$$

The corresponding maximum throughput $\gamma_{\max}$ can be computed from Eqs. 4.2 and 4.6 as

$$\gamma_{\max} = \frac{\lambda_0(1 + \lambda_0 t)}{2 + \lambda_0 t} \tag{4.7}$$

This proves the Theorem 1.

4.3.2 Discussion

Figure 4.3 shows plots of the throughput γ for varying levels of the incident traffic λ. The capacity [79] of the server μC and t are used as parameters. It can be seen that given a fixed μC, the served traffic increases as the threshold of delay time t increases. Also, for a fixed t, the served traffic increases as the service rate μC increases. Both these results are also intuitive. A larger service rate should obviously increase the throughput. Also, if there is a higher tolerance to delay, a higher level of throughput will have suffered delay lower than the specified threshold.

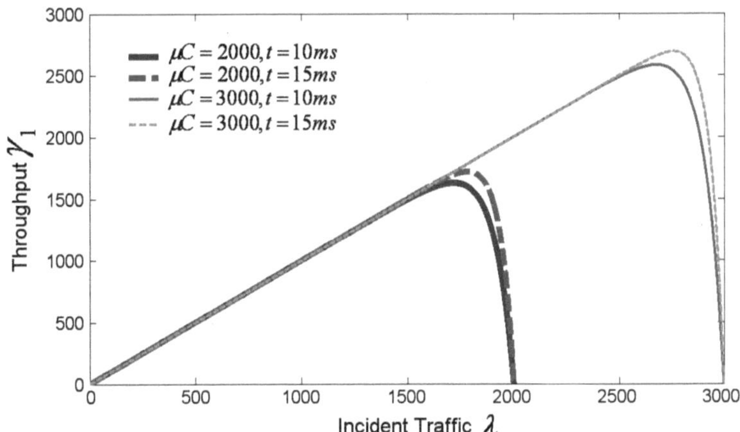

Fig. 4.3 Single-hop network throughput

Further, we note that the served traffic increases as λ increases, reaching a peak at $\gamma_{\max}$ and eventually declining to zero. It also follows that if the capacity of the network were fixed, the maximum throughput [80] for a given value of t can be computed and the maximum allowable incident traffic known. The relationship developed will allow sizing the resources needed against known requirements of threshold delay and incident traffic. The fact that the VoIP network should have a peak throughput for specified levels of threshold delay and service rate might warrant some further discussion. As the incident traffic increases from a low value, the delay suffered by packets will be initially small so the throughput will be equal to the incident traffic, thus constituting 100% throughput. As the incident traffic increases, the delay suffered by the packets will increase. This results in an increasing fraction of packets that will undergo delay larger than the threshold delay; this does not constitute effective throughput. For very large values of the incident traffic, the relative throughput will obviously approach zero.

It is instructive to compare the throughput performance illustrated in Fig. 4.3 with that of a corresponding M/M/1 system where no served traffic is discarded. In that case, the served traffic asymptotically reaches the server capacity. The decline of the served traffic to zero in our case is because as the incident traffic approaches the server capacity, the queuing time increases indefinitely resulting in the served traffic declining to zero. From Fig. 4.3, it can also be observed that as the server capacity increases, the throughput increases and, similarly, as the threshold delay increases, the throughput increases.

4.4 Two-Hop Tandem Network

In general, a VoIP call will go through a number of hops instead of a single-hop as shown in Fig. 4.2. In this section, we analyze a two-hop VoIP network.

A two-hop tandem network is shown in Fig. 4.4. Assume that the two links have the same capacity C. The analysis of throughput as a function of the threshold t and the incident traffic λ can be carried out as shown below.

The arrival process of the traffic incident on the second link can be assumed to be Poisson as well [58, 81]. In our analysis of the two-hop system, we assume that all traffic served by the first hop forms the incident traffic for the second hop, even if it was delayed beyond the threshold t. In other words, the policy of discarding

Fig. 4.4 A tandem VoIP
network

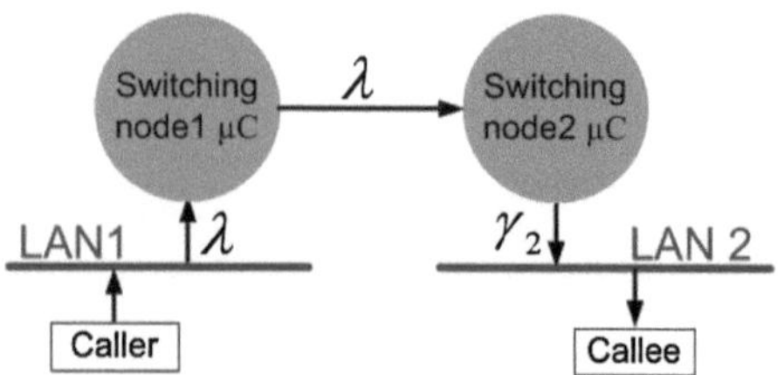

traffic with a delay higher than t is executed by the exit node. These two assumptions might want some further discussion. The Poisson character of the arriving messages at an intermediate node has been shown to be approximately correct through several measurements. The second assumption has been made in order to simplify the analysis. It will influence the result by marginally increasing the load on the network. The analysis will thus result in a slightly pessimistic value of the actual throughput.

The probability density function (PDF) of waiting time at the first node can de derived from (4.1) as:

$$f_{W_1}(t) = \mu C \rho (1 - \rho) e^{-\mu C(1-\rho)t} \tag{4.8}$$

A Laplace–Stieltjes Transform (LST) for the waiting time PDF $f_{W_1}(t)$ is given by

$$F(s) = \int_0^\infty f_{W_1}(t) e^{-st} dt = \rho \frac{\mu C(1 - \rho)}{s + \mu C(1 - \rho)} \tag{4.9}$$

The Laplace transform of the waiting time distribution for the two nodes in tandem, $F_C(s)$, can be now calculated as [82]:

$$F_c(s) = F(s) \times F(s) = \rho^2 \left[\frac{\mu C(1 - \rho)}{s + \mu C(1 - \rho)} \right]^2 \tag{4.10}$$

The waiting time pdf of the two-hop network can now be computed as:

$$f_c(t) = \rho^2 [\mu C(1 - \rho)]^2 t e^{-\mu C(1-\rho)t} \tag{4.11}$$

The relationship between pdf and tailend distribution is given by

$$F_c(t) = \int_0^t f_c(x) dx \tag{4.12}$$

From (4.11) and (4.12) we have

$$P(W_2 \leq t) = \rho^2 \{ 1 - e^{-\mu C(1-\rho)t} [1 + \mu C(1 - \rho)t] \} \tag{4.13}$$

where $P(W_2 \leq t)$ represents the probability distribution function of the total delay.

4.4.1 Analysis of Two-Hop VoIP Network

The maximum throughput of a two-hop M/M/1 system where each node is characterized by the same service rate μC and *the total threshold delay is t*, can now be developed as follows. We have

$$\gamma_2 = \lambda \cdot P(W_2 \leq t) = \lambda^3 \left(\frac{1}{\mu C}\right)^2 \left\{1 - e^{-(\mu C - \lambda)t}[1 + (\mu C - \lambda)t]\right\} \qquad (4.14)$$

From (4.14), we can get:

$$\frac{d\gamma_2}{d\lambda} = 3\lambda^2 \left(\frac{1}{\mu C}\right)^2 \left\{1 - e^{-(\mu C - \lambda)t}[1 + (\mu C - \lambda)t]\right\} - \lambda^3 \left(\frac{t}{\mu C}\right)^2 (\mu C - \lambda)e^{-(\mu C - \lambda)t}$$

$$(4.15)$$

The maximum value of γ_2 occurs when $\frac{d\gamma_2}{d\lambda} = 0$, or when λ_0 on the interval [0, μC] satisfies

$$3e^{(\mu C - \lambda_0)t} = 3 + (3 + \lambda_0 t)(\mu C - \lambda_0)t \qquad (4.16)$$

The maximum value of γ_2, or $\gamma_{2\mathrm{max}}$ can be derived from (4.14) and (4.16) as

$$\gamma_{2\,\mathrm{max}} = \frac{\lambda_0^4}{3}\left(\frac{t}{\mu C}\right)^2 (\mu C - \lambda_0)e^{-(\mu C - \lambda_0)t} \qquad (4.17)$$

4.4.2 Discussion

Figure 4.5 plots the throughput γ_2 for two different service rates and two different thresholds t. Relative to the single-hop network, one can readily observe the sharp decline in the network throughput if two links in tandem, each with an identical capacity, were to serve the same incident traffic while maintaining the same end-to-end threshold delay. By comparing the throughput performance, we note that the maximum achievable throughput of the two-hop VoIP network is consistently lower than that of the single-hop network for any specified level of threshold delay. Using the analytical results presented in this chapter, for specified values of the design parameters, namely the threshold delay and the incident traffic, the needed capacity of both the single-hop and the two-hop networks can be analytically evaluated. In the following analysis, we extend the results to an n-hop network.

4.5 Multiple-Hop Network

VoIP networks could have multiple hops. Each voice packet transmitted over the packet-switched IP network would, generally speaking, transit over multiple hops connected in series.

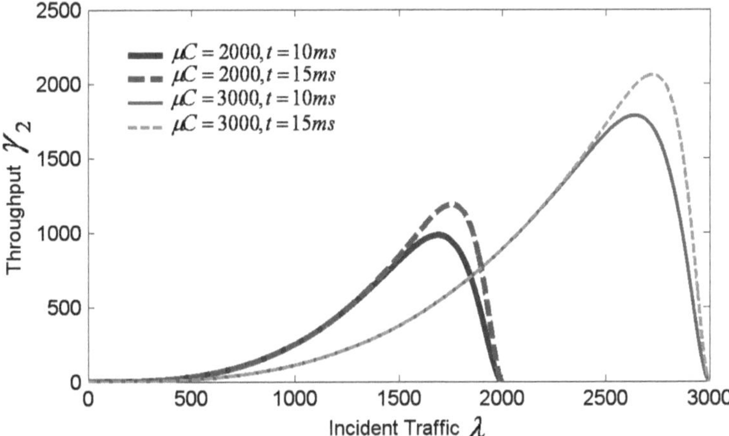

Fig. 4.5 Two-hop network throughput

4.5.1 Analysis of Multiple-Hop VoIP Network

As in the literature [62], the traffic is analyzed under the assumption that voice packets continue to follow the Poisson discipline at each intermediate node. Since we are interested in quantifying the impact of a multiplicity of hops, we can, without loss of generality, assume the server capacity and the arriving traffic at each node is identical. Only the last or the exit node drops the packets that have suffered delay higher than the threshold delay t. The impact of multiple hops on throughput is captured in Theorem 2.

Theorem 2 *The maximum throughput of a n-hop VoIP network, where each node is characterized as an individual M/M/1 system characterized with the same service rate μC and the threshold delay t, is given as:*

$$\gamma_{n\max} = \frac{\lambda_0^{n+2}}{(n-1)!}\left(\frac{t}{\mu C}\right)^n (\mu C - \lambda_0)e^{-(\mu C - \lambda_0)t}$$

This condition holds when the incident traffic λ_0 satisfies the following condition,

$$e^{(\mu C - \lambda_0)t} = \frac{\lambda_0 t[(\mu C - \lambda_0)t]^{n-1}}{(n+1)(n-1)!} + \sum_{k=1}^{n} \frac{[(\mu C - \lambda_0)t]^{n-k}}{(n-k)!}$$

Proof From (4.11), the density function $f_3(t)$ of the waiting time for the three-hop network can be obtained by convolution of the pdf associated with the first two-hop $f_2(t)$ and the third hop $f_1(t)$.

$$f_3(t) = \int_{-\infty}^{\infty} f_2(x)f_1(t-x)dx = \frac{1}{2}t^2[\mu C\rho(1-\rho)]^3 e^{-\mu C(1-\rho)t} \tag{4.18}$$

Continuing this convolution for each following hop, we can deduce the pdf for total waiting time in n-hop network:

$$f_n(t) = \frac{t^{n-1}}{(n-1)!}[\mu C\rho(1-\rho)]^n e^{-\mu C(1-\rho)t} \tag{4.19}$$

From (4.12) and (4.19) we have

$$P(W_n \leq t) = \rho^n \left(1 - e^{-\mu C(1-\rho)t} \sum_{k=1}^{n} \frac{[\mu C(1-\rho)t]^{n-k}}{(n-k)!}\right) \tag{4.20}$$

Therefore, the throughput for the n-hop network is given as:

$$\gamma_n = \lambda \cdot P(W_n \leq t) = \frac{\lambda^{n+1}}{(\mu C)^n}\left(1 - e^{-(\mu C - \lambda)t}\sum_{k=1}^{n}\frac{[(\mu C - \lambda)t]^{n-k}}{(n-k)!}\right) \tag{4.21}$$

Let $\frac{d\gamma_n}{d\lambda} = 0$, we then have:

$$e^{(\mu C - \lambda_0)t} = \frac{\lambda_0 t[(\mu C - \lambda_0)t]^{n-1}}{(n+1)(n-1)!} + \sum_{k=1}^{n}\frac{[(\mu C - \lambda_0)t]^{n-k}}{(n-k)!} \tag{4.22}$$

Moreover the maximum throughput of n-hop network can be obtained as:

$$\gamma_{n\,max} = \frac{\lambda_0^{n+2}}{(n-1)!}\left(\frac{t}{\mu C}\right)^n (\mu C - \lambda_0)e^{-(\mu C - \lambda_0)t} \tag{4.23}$$

This proves Theorem 2.

4.5.2 Discussion

Compared to the single-hop network which can achieve a hundred percent throughput if the upper delay bound goes to infinity, for the n-hop network, the maximum value of throughput can only reach ρ^n %.

4.6 Simulation Results

This section is intended to corroborate the analytical results presented above through the simulation of actual single-hop and multi-hop VoIP networks. The MATLAB simulator and ANSI C are used.

4.6.1 Simulation Scenario

In the platform, the system is considered at the network layer, i.e., the layer above the physical and the MAC layers. Each hop in VoIP networks is assumed to have a buffer with infinite memory. Packets waiting in the buffer are transmitted following the First-In-First-Out (FIFO) discipline. The incident traffic at each node is characterized by the Poisson distribution with the arrival rate λ. For M/M/1, the service rate follows negative exponential distribution with average value μ. The incoming packets are generated randomly and independently, and the probability of generating n packets at time interval t is given by

$$P_n(t) = \frac{e^{-\lambda t}(\lambda t)^n}{n!} \tag{4.24}$$

This is the Poisson distribution.

The time interval for the next generated packet is obtained using [83],

$$t = -\frac{1}{\lambda}\log(1-x) \tag{4.25}$$

where x is the uniform random number between 0 and 1.

At each simulation iteration, a packet is generated. The separation between two successive generated packets is determined by (4.25). On the other hand, the number of served packets within these time intervals is logged as well. Thus, the buffer status at each iteration is updated by adding 1 packet and subtracting the number of packets served within the time interval between the currently generated packet and the previous one. The simulation assumes unlimited buffer size. There is no overflow situation. The buffer underflow is taken into consideration. The simulation logs the buffer status (number of packets in the buffer) along with the simulation time and is used to calculate the delay statistics.

4.6.2 Simulation Results

Two scenarios have been considered. The first one is for a one-hop network and the other one is for a two-hop network with identical mean arrival rate λ at each buffer and identical mean service time $\frac{1}{\mu}$. Figure 4.6 shows the cumulative distribution function (CDF) of the waiting time in and Fig. 4.7 describes the single-hop system. Figures 4.8 and 4.9 describe the two-hop system. As shown in the figures, the simulation results closely match the analytical results derived in the previous sections.

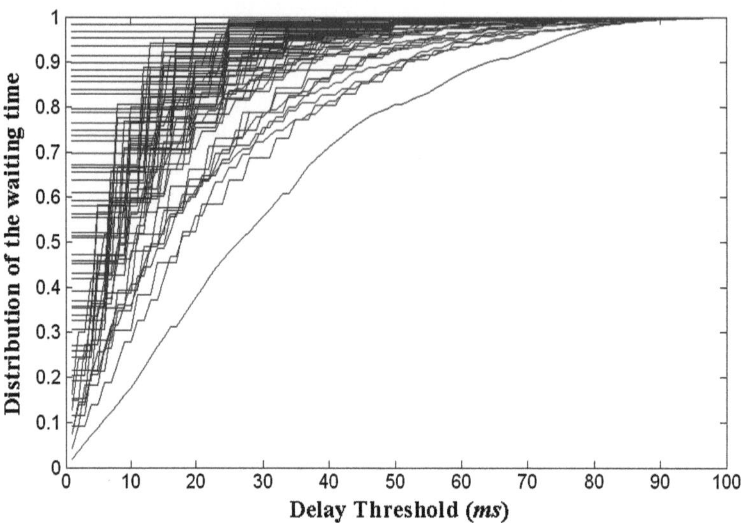

Fig. 4.6 Simulation of the delay distribution function of the single-hop

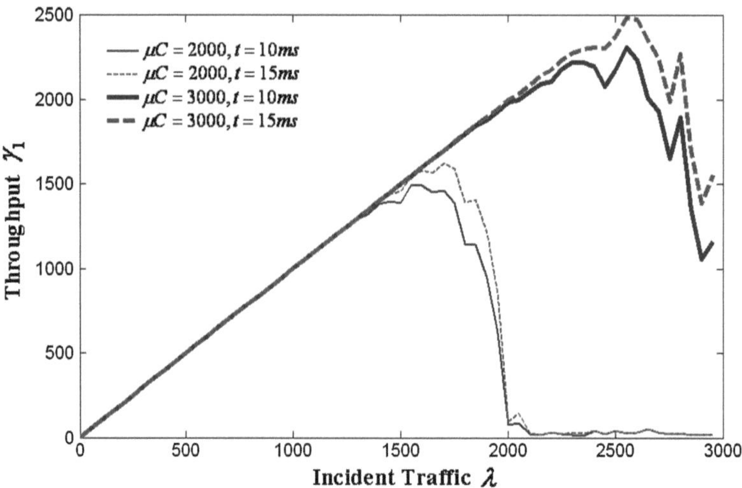

Fig. 4.7 Simulation of the throughput in the single-hop network

4.7 Conclusion

This chapter has presented a closed form solution relating the impact of bounding delays in a VoIP network to a specified limit. Our focus has been on computing throughput of a network if the traffic that suffered higher than a specified threshold delay did not constitute throughput. The results obtained will be useful in sizing up resources of the network to meet a specified maximum end-to-end delay criterion.

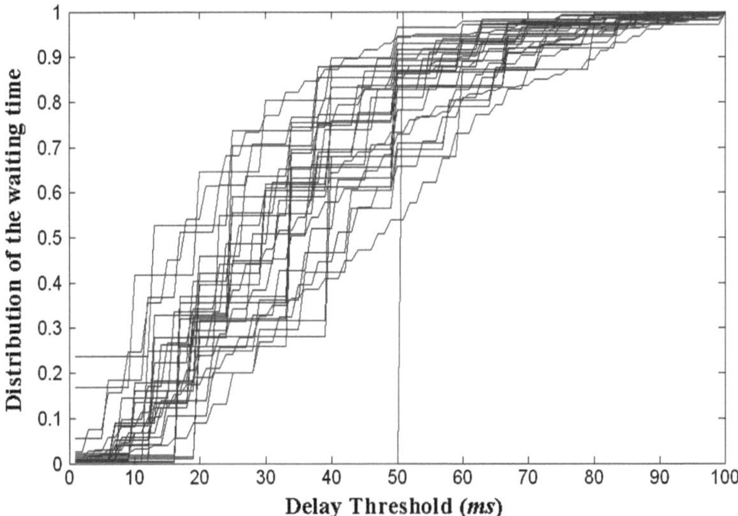

Fig. 4.8 Simulation of the delay distribution function of the two-hop

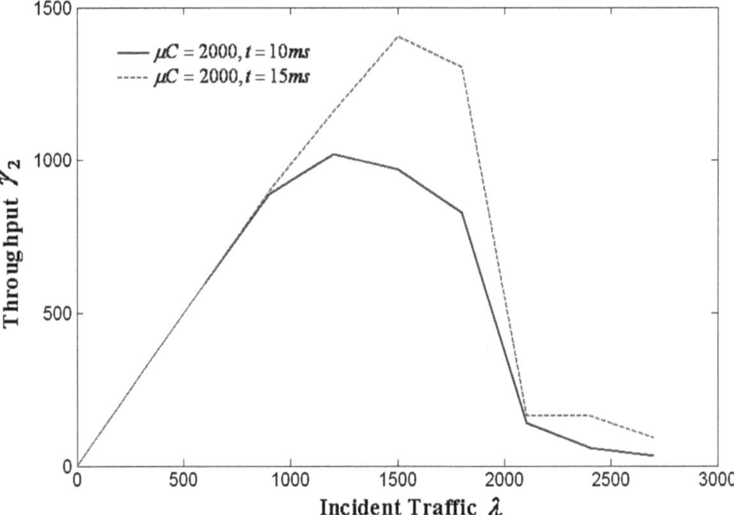

Fig. 4.9 Simulation of the throughput in the two-hop network

The results obtained can be used to compute the maximum traffic bearing capacity of the network. The negative impact of transiting through a large number of hops between the source and the destination can be readily observed. In the next chapter, we will provide the analytical solution and simulation performance when the voice packets are assumed to be of constant length. We will use results from queuing theory for the M/D/1 traffic discipline.

Chapter 5
Impact of Bounded Delay on Resource Consumption-M/D/1 Model

Abstract This chapter investigates the impact of bounded delays on throughput in VoIP networks modeled by M/D/1. Chapter 4 modeled the VoIP traffic as an M/M/1 system. Although the M/M/1 system is easier to analyze, the M/D/1 assumption for analyzing VoIP system might be more realistic in some cases. In contrast to contemporary literature that largely focuses on average delay to estimate the Quality of Service, our model focuses on an upper bound of delay, referred to as delay threshold. Traffic that exceeds the delay threshold is treated as lost throughput. The results obtained can be used in scaling resources in a multi-hop network for attaining specified levels of throughput under different thresholds of acceptable delays. Both single-hop and multi-hop transfers are addressed. The theoretical analysis presented in this chapter is further corroborated by simulation. The findings presented in this chapter will be very relevant to multi-hop network applications where received data that are older than a specified threshold period are not relevant and must be discarded. The contents of this chapter have partially been published in [84].

5.1 Introduction

The legacy circuit-switched network allocates dedicated bandwidth resulting in virtually no delay. In contrast to circuit- switched networks, packets in a packet-switched network undergo varying amounts of delay at each transit hop. Therefore, delay becomes a significant parameter in packet-switched networks. The delay addressed in this chapter is the variable queuing delay, the time each voice packet has to wait at each router in the path of the connection.

From the network designers' point of view, one major concern is how long it will take the packets to reach the destinations. If packets traverse multiple hops, the delay will increase. For real-time applications, a large delay will be unacceptable (ITU-T Recommendation G.114, 1996). As mentioned in Chap. 4, there

P. K. Verma and L. Wang, *Voice over IP Networks*, 49
Lecture Notes in Electrical Engineering, 71, DOI: 10.1007/978-3-642-14330-4_5,
© Springer-Verlag Berlin Heidelberg 2011

are two important parameters that capture the performance of a multi- hop network: the end-to-end delay and throughput. In contemporary literature on networks [85], the average delay has been usually used as a measure of the delay. However, most real-time applications would benefit from knowledge of the information conveyed that has an end-to-end delay below a defined threshold [86]. Accordingly, this chapter addresses the impact of acceptable upper delay bounds on throughput in multi-hop networks. We define effective throughput of the network as throughput that has suffered delay below a specified maximum. The findings presented in this chapter will be very relevant to multi-hop network applications where received data that are older than a specified threshold period are not relevant and must be discarded. The impact of multiple hops in the case of a circuit-switched network has been recently studied by one of the authors [18]. The results obtained for the case of circuit switched networks for the first time mathematically explained why and how the call completion rate of a hierarchical network would deteriorate as the number of hierarchies a call would have to travel increases (eventually approaching zero) in the case of a multi-hop network. The present work addresses packet-switched networks under the M/D/1 discipline. Since in a packet-switched network with infinite buffers no calls can be lost, the performance characterization is done through end-to-end delays. This is the subject of this chapter. The results obtained are compelling and offer a mirror image of the results previously obtained in the circuit-switched scenarios. The major focus of the work is on understanding the impact of multiple hops on throughput, e.g., when the data being transmitted are time sensitive and any delay in the received data beyond a specified value will render it useless. The transmission channel is assumed to be both lossless and contention-free. The analytical and simulation results shed light on just one fact: the cumulative delays in a multi-hop communication for time-sensitive data quickly render the received data useless as the number of hops increases.

The rest of the chapter is organized as follows:Sect. 5.2 addresses the queuing network model based on the M/D/1 system,Sects. 5.3, 5.4 and 5.5 present the analytical results for VoIP networks consisting of single hop, two hops and multiple hops, respectively, Sect. 5.6 confirms the results of the analytical solution by simulation, and Sect. 5.7 presents the conclusion of our work.

5.2 Network Model

Most analyses involving packet-switched systems assume that packets are negative exponentially distributed [87]. We assume a constant length for the size of each message. Accordingly, our model of the multi-hop network is based on M/D/1queues. An M/M/1 system is characterized by two parameters, the average packet arrival rate λ and the average service time $1/\mu$. An equivalent M/D/1 system would have a constant service time of $1/\mu$ and an arrival rate of λ. We use the parameter h to indicate the number of hops between the source and the destination.

Since VoIP networks are largely deployed to capture and transport formatted information, it is appropriate to assume a constant length for the size of each message. The M/D/1 model can be more appropriate for an IP network transporting voice where all voice codecs produce flows of packets of the same size, implying that all voice packets have the same deterministic service time [88]. The M/D/l model can be extended to multiple channels as the $M/(D_1 + D_2 + \cdots + D_n)/1$ model shown in [89, 90] for a VoIP network where the voice flows are produced by n types of codecs. This chapter will focus on the VoIP networks based on M/D/1 queues. Each of these queues is served using the FCFS (First Come First Serve) discipline.

We further assume that, for multi-hop traffic, the traffic at each transit node continues to follow the Poisson model for the arrival process [91, 92].

5.3 A Single-Hop Network

We first consider a single-hop system, where the packets are transmitted from the source node to one of the adjacent nodes.

The distribution function of the waiting time W_1 in a single-hop M/D/1 system is given by [93],

$$p_1 = P(W_1 \leq t) = (1 - \lambda) \sum_{j=0}^{T} \frac{[\lambda(j - t)]^j}{j!} e^{-\lambda(j-t)}, \tag{5.1}$$

where $t = T + \tau$, i.e., $T = \lfloor t \rfloor$, where $\lfloor t \rfloor$ is the largest integer less than or equal to t.

Therefore, the (residual) throughput of the single-hop network, i.e., the throughput that only includes packets with a queuing delay not exceeding the threshold delay, can be expressed as:

$$\gamma_1 = \lambda p_1, \tag{5.2}$$

where λ is the arrival rate of packets.

The normalized throughput, i.e., throughput expressed as a function of the incident traffic can be given as:

$$p_1 = \frac{\gamma_1}{\lambda} \tag{5.3}$$

Figure 5.1 shows the waiting time distribution for M/D/1 system with different values of λ. The mean service time (which represents the length of packets) is 1 in each case. As shown in the Fig. 5.1, the normalized throughput increases with an increase in the threshold delay, reaching 100% asymptotically. We also note from Fig. 5.1 that the M/D/1 system consistently shows a higher normalized throughput for a smaller λ or, equivalently, smaller ρ.

Using Eqs. 5.2 and 5.3, Fig. 5.2 depicts the throughput γ_1 as a function of the incident traffic λ with several values of the threshold delay t as a parameter. It can

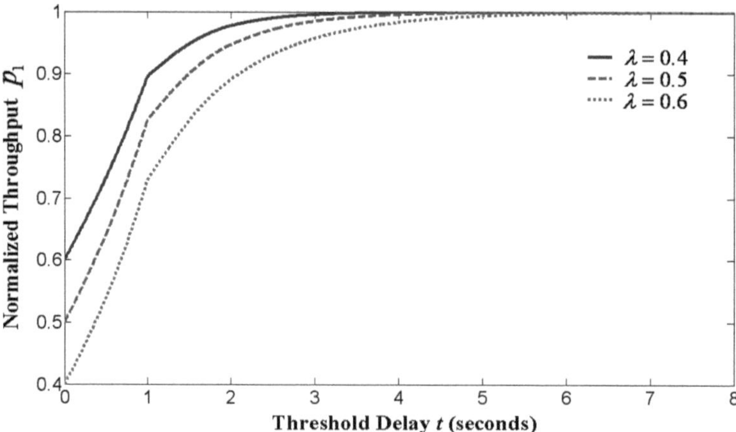

Fig. 5.1 The waiting time distribution of a single-hop network

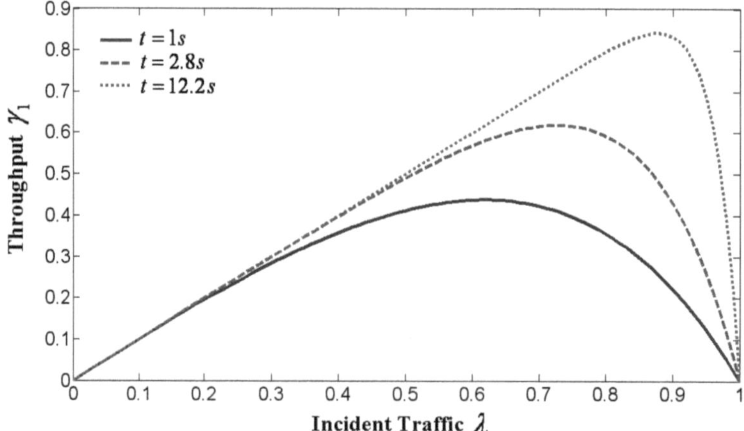

Fig. 5.2 The throughput of a single-hop network (M/D/1)

be seen that, given the same incident traffic λ, the throughput increases as the threshold delay t increases. It is also interesting to find out that, for each threshold delay t, the throughput initially increases with λ, reaching a peak γ_{max}, and eventually declines to zero. In other words, for a given t, the throughput is maximized for a specific value of λ. Unfortunately, due to the nature of Eq. 5.1, the specific value of λ corresponding to the maximum value of throughput cannot be evaluated in an explicit form. However, it can be numerically evaluated indicating that, for a pre-specified threshold delay, the throughput is maximized at a specific level of incident traffic that can be numerically derived. The interplay among the threshold delay, throughput and the incident traffic can be used in order to design a cost efficient VoIP network that ideally meets the specified needs of the user. We also note from Fig. 5.2 that the throughput falls much more sharply as λ increases, for higher value of t. The design parameters of a VoIP network thus have a very high sensitivity to λ for higher value of t.

The results depicted in Fig. 5.2 are somewhat anti-intuitive, although not difficult to comprehend under the stipulations of this investigation. If all served packets, irrespective of delay, were acceptable, the throughput would not show a decline and would reach 100% asymptotically. However, if the packets that suffered a delay higher than t were to be discarded and did not constitute throughput, the residual throughput would decline because a larger fraction of the packets would suffer delays higher than the threshold delay as the utilization factor (or the incident traffic) increased. The existence of a peak at which the VoIP network would deliver the maximum throughput is an important parameter that can be used to size the resources of the network.

5.4 Two-Hop Tandem Network

In general, packets will go through a number of hops instead of a single hop. We first consider a two-hop tandem network. We assume identical service time distribution and arrival process for each node [58]. For purposes of analysis and insight into design, but without any loss of generality, we also assume that the first node does not drop any packets, even though it might have exceeded the delay threshold t. The exit node will drop any packets that have cumulatively suffered a total delay higher than t.

The probability density function (PDF) of waiting time at the first node can be derived from Eq. 5.1,

$$f_1(t) = (1 - \lambda)\left[\sum_{j=0}^{T} \frac{\lambda^{j+1}}{j!}(j - t)^j e^{-\lambda(j-t)} - \sum_{j=0}^{T} \frac{\lambda^j}{(j-1)!}(j - t)^{j-1} e^{-\lambda(j-t)}\right]. \quad (5.4)$$

A LST (Laplace–Stieltjes Transform) for the waiting time PDF is given by

$$F_1(s) = \int_0^\infty f_{W_1}(t)e^{-st}dt = (1 - \lambda)\sum_{j=0}^{T}(-\lambda)^j e^{-js}\left[\frac{\lambda}{(s - \lambda)^{j+1}} + \frac{1}{(s - \lambda)^j}\right]. \quad (5.5)$$

The Laplace transform of the total (combined) waiting time PDF is equal to the product of the Laplace transforms of the waiting time PDF associated with each node [82]. Thus,

$$F_2(s) = F_1^2(s)$$
$$= (1 - \lambda)^2\left\{\sum_{j=0}^{T}(-\lambda)^j e^{-js}(j + 1)\left[\frac{1}{(s - \lambda)^j} + \frac{2\lambda}{(s - \lambda)^{j+1}} + \frac{\lambda^2}{(s - \lambda)^{j+2}}\right]\right\}.$$
$$(5.6)$$

The time-domain PDF of the two-node network can be found as:

$$f_2(t) =$$

$$(1 - \lambda)^2 \sum_{j=0}^{T} \frac{(-\lambda)^j}{j!} \Big\{ j(j+1)(t-j)^{j-1} + 2\lambda(j+1)(t-j)^j + \lambda^2(t-j)^{j+1} \Big\} e^{\lambda(t-j)}.$$

$$(5.7)$$

The relation between the PDF and tail end distribution is given by

$$F_c(t) = \int_0^t f_c(x)dx. \tag{5.8}$$

From (5.7) and (5.8) we have,

$$p_2 = P(W_2 \le t) = (1 - \lambda)^2 \sum_{j=0}^{T} \frac{1}{j!} \Big\{ (j+1)[\lambda(j-t)]^j - [\lambda(j-t)]^{j+1} \Big\} e^{-\lambda(j-t)} \tag{5.9}$$

where p_2 is the normalized throughput of a tandem (two-hop) system, and we have

$$\gamma_2 = \lambda p_2 \tag{5.10}$$

Figure 5.3 depicts the normalized throughput as a function of the threshold delay t with λ as a parameter. It can be seen that given the same threshold delay, the normalized throughput of the two-hop system is higher for a smaller λ. In other words, a higher fraction of traffic suffers delay not exceeding the threshold delay under light low conditions. (As the threshold delay increases, the residual throughput can, however, increase even though a smaller fraction of packets constitute the residual throughput.) It can also be seen from the distribution of the waiting time in a two-hop M/D/1 system, that the normalized throughput p_2 can reach a maximum value of 100% as the threshold delay increases to infinity.

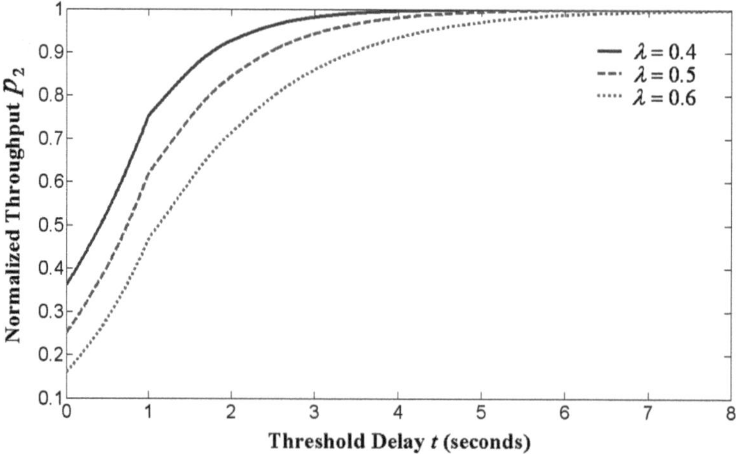

Fig. 5.3 The waiting time distribution of a two-hop network

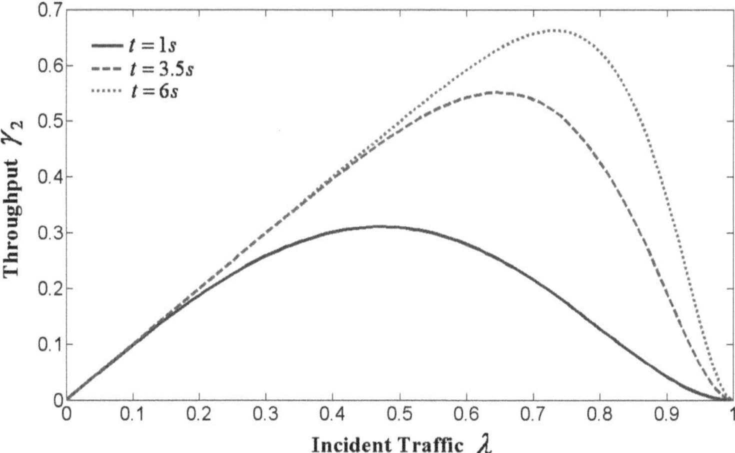

Fig. 5.4 The throughput of a tandem network as a function of the incident traffic

Figure 5.4 shows plots of the throughput γ_2 for varying levels of the incident traffic λ. The threshold delay t is used as the parameter. It can be seen that, as in the single-hop network case, for a given λ, the throughput is always higher (or asymptotically equal to) for a higher value of the threshold delay. Also, and again similar to the single-hop network, we can readily observe that for a given end-to-end threshold delay, the throughput increases as λ increases, until it reaches a maximum, and then declines to zero when the incident traffic approaches the service rate, i.e., when $\rho = 1$. Both $\gamma_{2\text{max}}$ and the corresponding λ can be numerically calculated from Eqs. 5.9 and 5.10.

A comparative visualization of the single and multi-hop VoIP network performance is appropriate at this point. By comparing the throughput performance, we note that the maximum achievable throughput of the two-hop VoIP network is consistently lower than that of the single-hop network for any specified level of threshold delay. Using the analytical results presented in this chapter, for specified values of the design parameters, namely the threshold delay and the incident traffic, the needed capacity of both the single-hop and the two-hop networks can be analytically evaluated. The additional transmission capacity required in the two-hop case can then be compared against, for example, the higher power requirement of the single-hop case since not all the nodes can access each other with the relatively low power needed for a two-hop network.

5.5 Multi-Hop Network

VoIP networks could have multiple nodes for reasons cited earlier. In such networks, each packet transmitted over the VoIP can, generally speaking, transit over multiple hops connected in series. As in the two-hop case, we analyze the throughput under the assumption that packets continue to follow the M/D/1 discipline at each intermediate node [62]. Furthermore, only the last or the exit node drops the packets that have suffered a cumulative queuing delay higher than the threshold delay t.

The end-to-end queuing delay in a multi-hop network is the sum of the waiting times in each hop along the series of h-hop [94]. Since we assume constant service time for each packet, we have

$$W_h = \underbrace{W_1 + W_1 + \cdots + W_1}_{h} = h \times W_1. \tag{5.11}$$

The PDF of the end-to-end delay $f_h(t)$ can be derived from the convolution of the PDF of the waiting time in each hop, i.e.,

$$f_h(t) = \underbrace{f_1(t) \otimes f_1(t) \otimes \cdots \otimes f_1(t)}_{h}. \tag{5.12}$$

The LST of the convolution will be the multiplication of the LST at each hop, i.e.,

$$F_h(s) = \underbrace{F_1(s) \times F_1(s) \times \cdots \times F_1(s)}_{h} = F_1^h(s). \tag{5.13}$$

By inverting the LST in (5.13), the end-to-end distribution will be obtained from (5.8)

$$p_h = P(W_h \le t) = (1-\lambda)^h \sum_{j=0}^{T} \sum_{i=0}^{h-1} \frac{(-1)^i}{i!j!} \binom{h+j-1}{i+j} [\lambda(j-t)]^{i+j} e^{-\lambda(j-t)}. \tag{5.14}$$

Using Eq. 5.14, one can evaluate the throughput of a multi-hop VoIP network where packets suffering an end-to-end delay higher than t are discarded as:

$$\gamma_h = \lambda p_h \tag{5.15}$$

As in the single-hop or multi-hop cases, given a threshold delay, the optimum capacity of each node can be determined. Alternatively, given a network of specified capacity, its effective throughput for any specified threshold delay can be numerically evaluated. The throughput performance of the multi-hop network continues to deteriorate as the number of hops increases. As in the case of the two-hop network, a trade off among the performance parameters and the resource requirements can be made leading to an informed choice among the topologies under consideration.

5.6 Simulation Results

This section is intended to corroborate the analytical results presented above through the simulation of actual single-hop and multi-hop networks [95]. We have used the MATLAB to simulate the multi-hop communication for different scenarios.

5.6.1 Simulation Scenarios

In our simulation, each node is modeled as a First-In-First-Out (FIFO) buffer with a limited size ($B_{\max}$). The incident traffic at each node is characterized by the

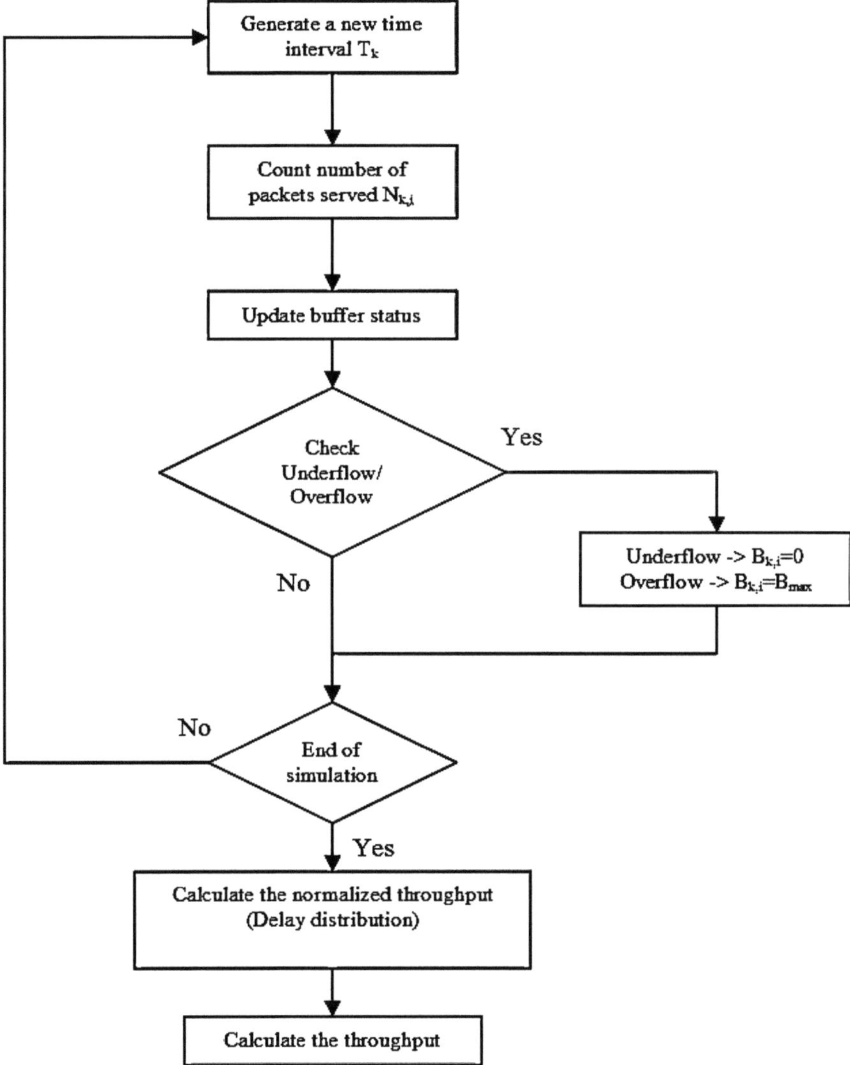

Fig. 5.5 Block diagram of the simulation

Poisson distribution with the arrival rate λ. For M/D/1, the service rate is fixed and is equal to μ.

Figure 5.5 is a block diagram of the simulation technique used to obtain the results. The simulation is based on variable simulation time steps, which are the time intervals between the arrival packets. At each simulation step, the buffer status at each node is updated. At the end of the simulation, the buffer status for each node is used to calculate the following performance metrics:

1. The normalized throughput for different delay thresholds.
2. The network throughput for different values of arrival rates.

The meanings of the symbols used in Fig. 5.5 are as follows: T_k is the kth simulation time step; N_{ki} the number of packets served at node i in a duration of T_k; B_{ki} the number of packets in the buffer of node i (buffer status of node i) at the kth simulation step; and $B_{\max,i}$ the buffer size for node i.

At the beginning of the simulation, it is assumed that the buffers are empty. During each iteration, the program calculates the next time interval to generate a new packet, which is assumed to have fixed length. The time intervals between generated packets are random and independent since they follow Poisson distribution. Since the probability of generating n packets in time interval t is given by

$$P_n(t) = \frac{e^{-\lambda t}(\lambda t)^n}{n!}. \tag{5.16}$$

Therefore, we can find the time interval for the next generated packet as [83],

$$t = -\frac{1}{\lambda}\log(1 - x) \tag{5.17}$$

where x is the uniformly distributed random number between 0 and 1.

Equation 5.17 is used in the simulation to calculate the next time interval between generated packets. As shown in the block diagram, at iteration k, the next time interval before generating a new packet is T_k. Then the program calculates the number of packets served at each node during T_k.

The simulation logs the buffer status (number of packets in the buffer) along with the simulation time intervals. These values are used to calculate the throughput statistics. Figure 5.6 shows the buffer status of one node during one simulation iteration. Since the simulation assumed a fixed packet size, it is justifiable to consider the delay of each packet in the node's buffer as being proportional to the number of packets queued in the buffer.

At the end of the simulation, the sequence of time intervals and buffer status at each node are used to calculate the histogram of the buffer status at each node. The normalized version of this histogram is observed as a probability distribution of packet delays at each node. This observation is valid since the serving time and the packet length values are fixed. The cumulative probability distribution is obtained by integrating the probability distribution of packet delay. This cumulative

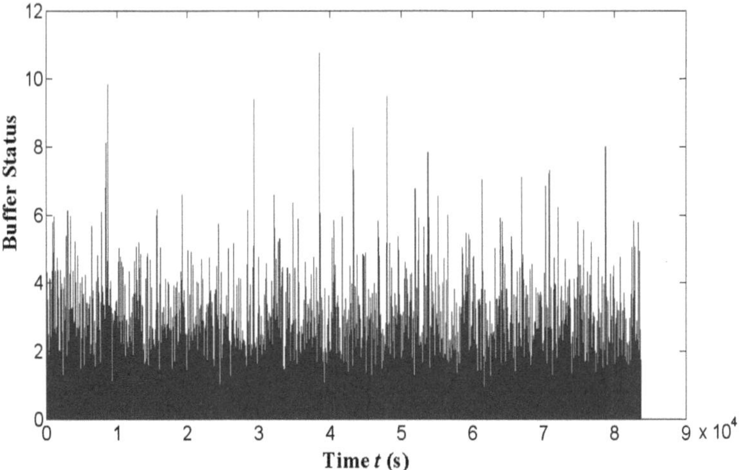

Fig. 5.6 Buffer status

probability distribution represents the normalized throughput for different delay thresholds. The simulation is repeated for different values of arrival rates λ to obtain different cumulative probability distributions from which the throughput for different arrival rates can be calculated.

Two scenarios have been considered, the first one is for one-hop networks and the other is for two-hop networks with identical arrival rate λ for each buffer and identical service time $1/\mu$. The simulation results are found to match closely the analytical results derived in the previous sections.

5.6.2 One-Hop Network Simulation Results

The simulation is performed for a one-hop network by assuming a single source of packet generation with a Poisson distributed arrival. The simulation iteration stops after generating 10,000 packets and is repeated for 100 different normalized values of arrival rates l varying from 0 to 1. Figure 5.7 shows the normalized throughput at different delay thresholds for three different values of l namely: 0.4, 0.5 and 0.6. This figure is obtained by integrating the normalized histogram of the node buffer status, which is the packet-generating node in the case of the one-hop network scenario.

Figure 5.8 represents the throughput of the network for different values of λ and is obtained for three different delay thresholds, which are 1, 2.8, and 12.2 s. This plot is generated using the cumulative probability value (normalized throughput) at each of the three delay thresholds (which can be found using Fig. 5.7) multiplied by the corresponding value of λ.

The simulation results can produce smoother lines by increasing the number of packets generated for each iteration and for more values of λ in the range between

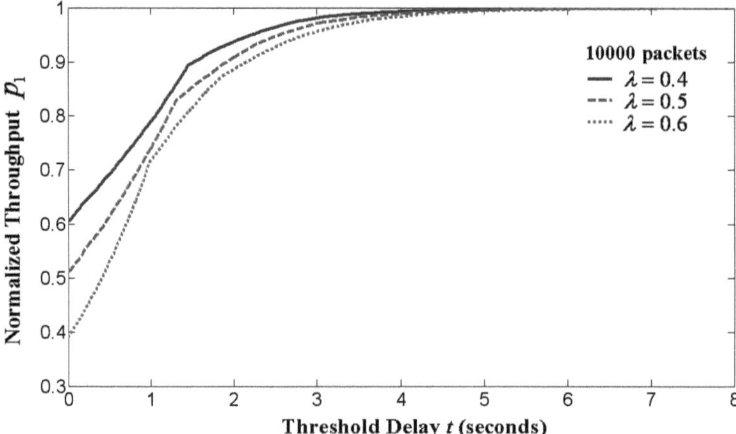

Fig. 5.7 Simulation result of the waiting time distribution

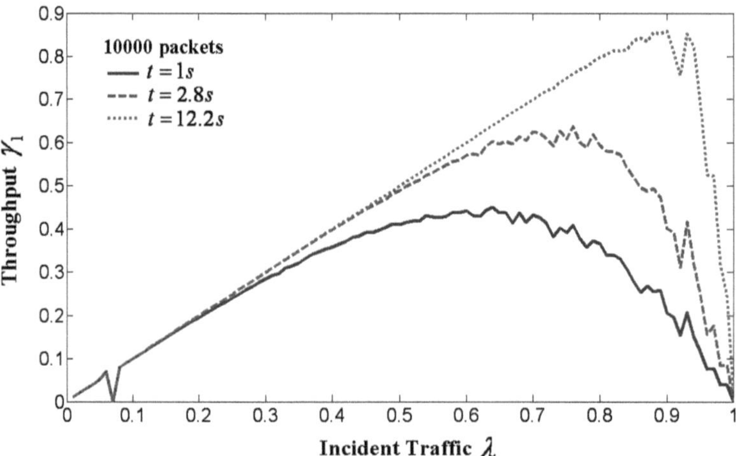

Fig. 5.8 Simulation result of the throughput for the single-hop network

0 and 1. However, the values 10,000 for the number of generated packets and 100 for the number of values of the arrival rate λ between 0 and 1 were chosen to perform the simulation in a reasonable time.

5.6.3 Two-Hop Network Simulation Results

The two-hop simulation results are obtained by monitoring the status of two buffers for each node along the two-hop network. The first buffer belongs to the

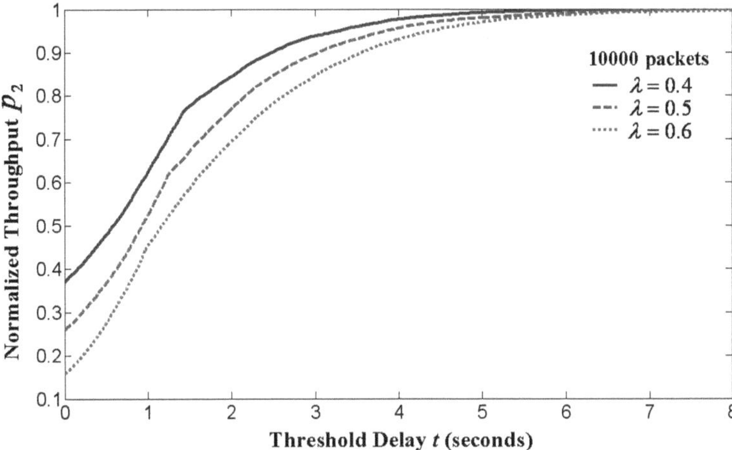

Fig. 5.9 Simulation result of the waiting time distribution

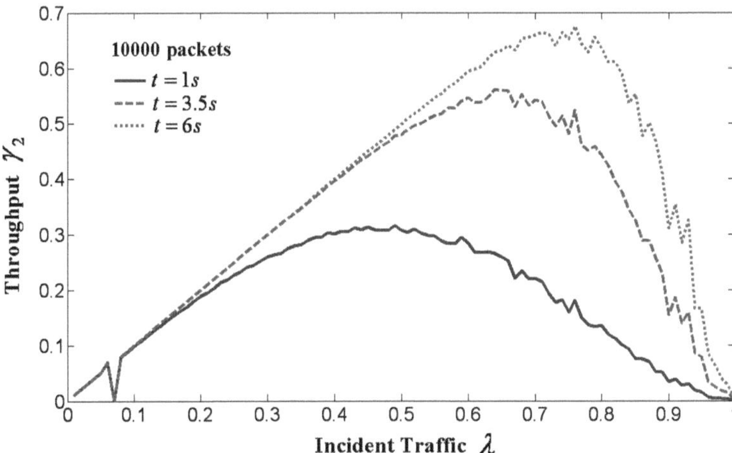

Fig. 5.10 Simulation result of the throughput for the two-hop network

packet- generating node while the second buffer is for the forwarding node between the generating node and the final destination node. Each buffer receives packets at a rate that is Poisson distributed, and the packets are served at a constant rate.

The simulation is repeated for 100 different normalized values of λ between 0 and 1. Figure 5.9 shows the normalized throughput at packet delay thresholds for three different values of λ namely 0.4, 0.5 and 0.6. The delay experienced by each packet is proportional to the total number of packets in the buffer of the packet generating node at the packet generation time instant plus the number of packets in the buffer of the forwarding node at the time instant of serving the packet at the generating node. Each simulation iteration stops after generating 10,000 packets and the delay experienced by each packet is calculated. Figure 5.8 is obtained by integrating the normalized histogram of 10,000 packet delays experienced by passing through two FIFO buffers.

Figure 5.10 represents the throughput of the network for different values of λ and is obtained for three different delay thresholds, which are 1, 2.8 and 12.2 s. This plot is generated in the same manner described for the one-hop network scenario.

5.7 Conclusion

This chapter has presented a closed form solution relating the impact of bounded delays on throughput in VoIP networks modeled by M/D/1. Our focus has been on computing throughput of a network if the traffic that suffered higher than a specified threshold delay did not constitute throughput. The results obtained will be useful in sizing up resources of the network in order to meet a specific throughput requirement with a specified maximum end-to-end delay criterion. The results obtained can be used to compute the maximum traffic bearing capacity of a network or in the design of a VoIP network with optimum throughput if the performance parameters and resource constraints were specified. The negative impact of transiting through a large number of hops between the source and the destination in terms of reduced throughput can be readily observed. The analytical results derived in this chapter have been further corroborated by simulating example networks and comparing the simulation results with the analytical results.

Chapter 6
Impact of Bounded Jitter on Resource Consumption in Multi-Hop Networks

Abstract Jitter is, potentially, the largest source of degradation in the quality of voice in VoIP systems. This chapter presents an analytical solution to the following question: How is jitter affected by the number of hops that VoIP packets travel over and, if the end-to-end jitter were to be bounded to a pre-defined value, how would the resources in the network need to be scaled up as the number of hops increases? The chapter also provides a way to compute the traffic handling capability of a multi-hop resource-constrained network under a defined limit of end-to-end jitter.

6.1 Introduction

Jitter is a potential source of quality degradation that can considerably reduce the QoS of VoIP communication. Among the parameters that define end-to-end performance, jitter is potentially the most significant. Accordingly, its containment is a major factor in the design of VoIP networks [96, 97]. Jitter is characterized in terms of the variance of the interval between successive packets at the receiving end *relative to that at the transmit end*. Hence, a reasonable measure of jitter in VoIP systems is in terms of the variance of the packet delay [98, 99]. In this chapter, we evaluate the variance of the delay for both single-hop and multi-hop networks. The primary objective of this chapter is to present an analytical solution that answers the following question: How is jitter impacted by the number of hops that VoIP packets travel over and, if the end-to-end jitter were to be bounded to a pre-defined value, how would the resources in the network need to be scaled up as the number of hops increases? The chapter also provides a way to compute the traffic handling capability of a multi-hop resource-constrained network under a defined limit of end-to-end jitter. Since our focus is on relating the impact of the number of hops to QoS degradation, we make several simplifying assumptions

P. K. Verma and L. Wang, *Voice over IP Networks*,   63
Lecture Notes in Electrical Engineering, 71, DOI: 10.1007/978-3-642-14330-4_6,
© Springer-Verlag Berlin Heidelberg 2011

without sacrificing the generality of our findings. We apply a Markov structure in deriving the jitter steady-state statistics for the multiplexed un-correlated traffic [100, 101]. We assume that the buffer capacity at each node is infinite.

The chapter is organized as follows. Section 6.2 presents the assumptions, and generalizes the results to a two-hop and then to a multi-hop network. It also derives jitter as a function of traffic statistics in a single-hop network, i.e., when the source and the destination are separated by a single router. Section 6.3 analyzes the impact of the number of hops from a variety of perspectives. Section 6.4 presents our conclusions.

6.2 Jitter Analysis

6.2.1 Single-Hop Model

We first consider voice traffic served by a single server which is our model for a single-hop network. We model the arriving VoIP packets as M/M/1 traffic [76–78].

In this model, the jitter can be estimated by the variance of the delay σ_D^2 [102], given as

$$\sigma_D^2 = \frac{1}{\mu^2 C^2 (1 - \rho)^2}. \tag{6.1}$$

The following section addresses how jitter accumulates as the number of hops increases.

6.2.2 Two-Hop and Multi-Hop Model

Figure 6.1 shows single and multi-hop networks graphically. As in the literature [62], we compute the variance of the delay under the assumption that voice packets continue to follow the Poisson discipline for the arrival process at each intermediate hop. Since we are primarily interested in quantifying the impact of a multiplicity of hops on the accumulated jitter, we can, without loss of generality, assume that the server capacity and the traffic at each hop are identical. For the two-hop model, we have,

$$D = D_1 + D_2 \tag{6.2}$$

where D_1 and D_2 are the packet delays introduced by the first and the second server, respectively. Further, since

$$\sigma_D^2 = D^{(2)} - (D^{(1)})^2 \tag{6.3}$$

and since the delay at each hop D_1 and D_2 are independent of each other, we can write, for the two-hop system

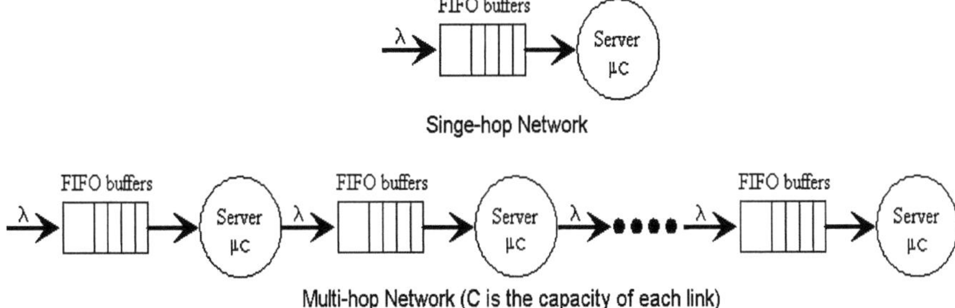

Fig. 6.1 Single-hop and multi-hop networks

$$\sigma_D^2 = [D_1^{(2)} - (D_1^{(1)})^2] + [D_2^{(2)} - (D_2^{(1)})^2] = \sigma_{D_1}^2 + \sigma_{D_2}^2 \tag{6.4}$$

which is the resultant jitter of a two-hop system.

Due to the assumption of identical traffic at each hop, the end-to-end jitter of packets that have traversed n hops can be obtained as,

$$\sigma_D^2 = \sum_{i=1}^{n} \sigma_{D_i}^2 = n\sigma_{D_1}^2. \tag{6.5}$$

This shows that the delay variance, and hence the jitter of the single-hop network, is lower by a factor of n compared to the n-hop network with the same bandwidth at each hop, and identical incident traffic at each node or server. We now investigate how the accumulated jitter affects the capacity needed for a multi-hop network, if the jitter were to be equal to a corresponding single-hop system.

From (6.1) and (6.5), we can express the jitter of a multi-hop network as,

$$\sigma_D^2 = \frac{n}{\mu^2 C^2 (1 - \rho)^2}. \tag{6.6}$$

Section 6.3 presents the cumulative impact of the number of hops on jitter in graphical form.

6.3 Impact of the Number of Hops on Jitter

This section presents the impact of the findings in Sect. 6.2 on the end-to-end jitter from a number of perspectives.

6.3.1 Capacity Requirement for a n-Hop Network as a Function of n with a Pre-Defined Jitter Upper Bound

From Eq. (6.6), since $\rho = \lambda/\mu C$, we can write, after some algebraic simplification,

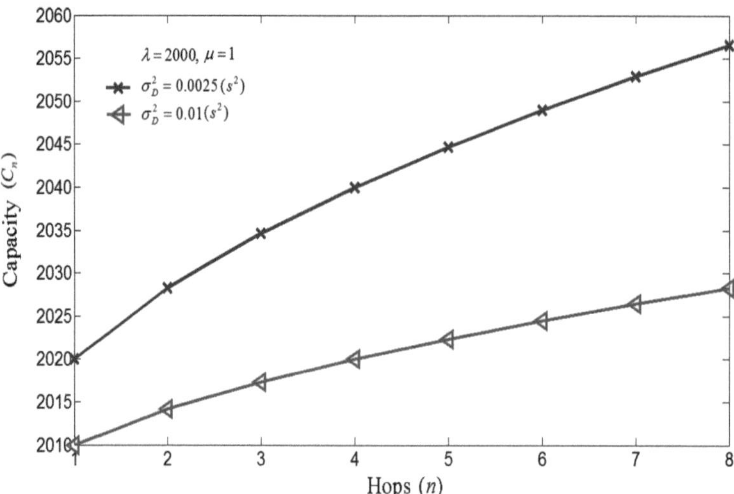

Fig. 6.2 Capacity required as a function of n with jitter as a parameter

$$C = \frac{1}{\mu}\left(\lambda + \sqrt{\frac{n}{\sigma_D^2}}\right). \tag{6.7}$$

Equation (6.7) shows that if the end-to-end jitter were to be fixed to σ_D^2, and the traffic parameters are given as λ and μ, the capacity needed at each hop can be evaluated as a function of the number of hops, n. Figure 6.2 depicts the relationship graphically. The traffic parameters used are $\lambda = 2,000$ and $\mu = 1$. The two jitter bounds used as parameters are, $\sigma_D^2 = 0.0025\,(s^2)$ and $\sigma_D^2 = 0.01\,(s^2)$. It can be observed from Fig. 6.2 that the capacity needed *for each hop* in n-hop network increases as the pre-defined jitter upper bound decreases. For example, for $n = 1$, the capacity needed is 2,010 bps for a jitter bound $\sigma_D^2 = 0.01\,(s^2)$ while we need a capacity of 2,020 bps if the jitter bound were to be reduced to $\sigma_D^2 = 0.0025\,(s^2)$, a difference of 10 bps. The corresponding figures for $n = 4$, require double the additional capacity (20 bps) to maintain the jitter bound to the lower value of $\sigma_D^2 = 0.0025\,(s^2)$.

6.3.2 The Impact of the Utilization Factor on the Capacity per Hop Needed for a Pre-Defined Upper Bound on End-to-End Jitter

In this section, we will focus on the impact of the number of hops on the capacity needed using ρ as the parameter while μ and λ are kept constant. Equation (6.6) can be rearranged to yield,

$$C = \frac{1}{\mu(1-\rho)}\sqrt{\frac{n}{\sigma_D^2}} \tag{6.8}$$

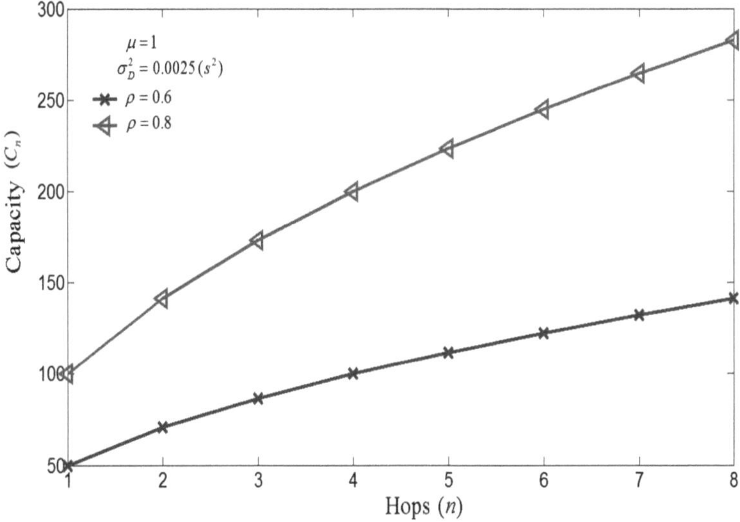

Fig. 6.3 Capacity required as a function of n with different utilizations

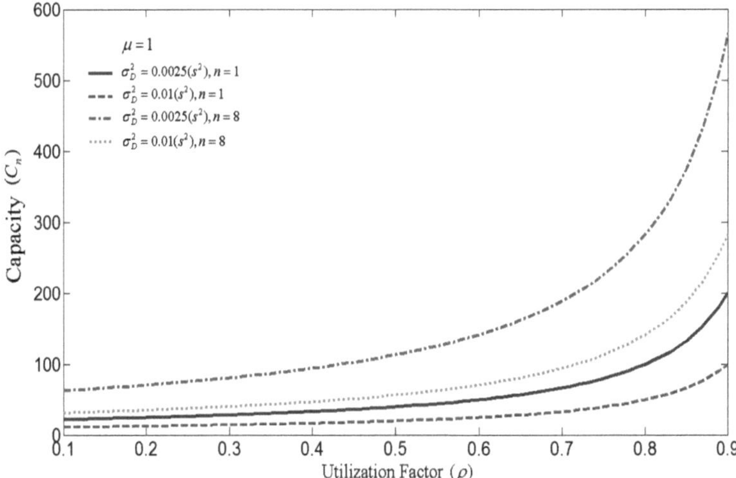

Fig. 6.4 Capacity requirements as a function of ρ, with n and σ_D^2 used as parameters

Figure 6.3 shows the results graphically. In particular, it illustrates that the capacity needed increases dramatically as the utilization factor increases. For example, the capacity needed at each hop in an eight-hop network is more than twice if the same traffic were to traverse a single hop and if the end-to-end jitter were to remain identical in each case.

Figure 6.4 focuses on how the capacity needed varies as a function of the utilization factor. As shown previously, the smaller the value of the jitter upper bound, the more the capacity needed at each hop. In addition, the capacity needed at each hop increases slowly initially, but the increase is much more rapid as the number of hops or as the utilization factor increases. The following section addresses the impact on throughput in a resource-constrained network if the jitter were to be bounded to an upper value.

6.3.3 The Impact on Throughput of a Resource-Constrained Multi-Hop Network with a Pre-Defined Upper Bound on End-to-End Jitter

In the case of a resource-constrained network, where the transmission capacity for each hop is limited to a pre-defined value C, the end-to-end jitter can be controlled to remain within a pre-defined upper bound, if the traffic is reduced as the number of hops n increases. Again, from Eq. (6.6), we can write:

$$\lambda = \mu C - \sqrt{\frac{n}{\sigma_{\mathrm{D}}^2}}. \tag{6.9}$$

If the end-to-end jitter were limited to σ_{D}^2, then for $\mu C = 2,000$, the capacity needed can be evaluated as a function of n. The two jitter bounds used are the same as in Fig. 6.2, $\sigma_{\mathrm{D}}^2 = 0.0025(s^2)$ and $\sigma_{\mathrm{D}}^2 = 0.01(s^2)$. The results are graphically depicted in Fig. 6.5.

It can be readily observed that, in order to keep the end-to-end jitter bounded to $0.01s^2$, the incident traffic must be reduced from 1,990 packets per second in a

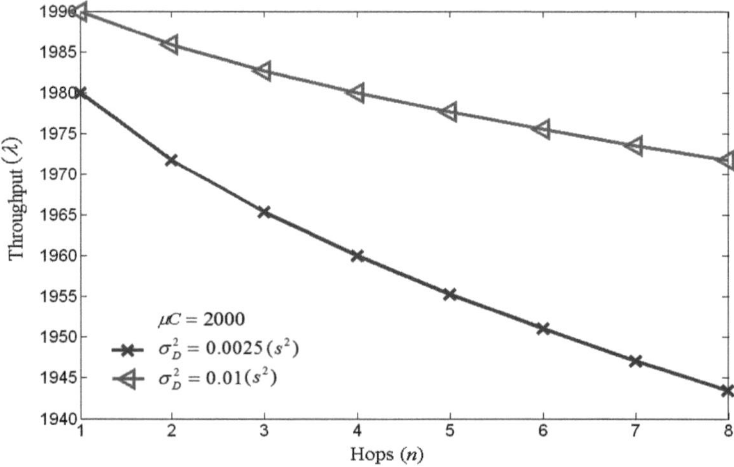

Fig. 6.5 Throughput of the n-hop network under an upper bound jitter and specified transmission capacity C at each hop

single-hop network to 1,980 in the four-hop network. Similarly, for a lower end-to-end jitter bound of $0.0025s^2$, the incident traffic must be reduced from 1,980 packets per second in the single-hop network to 1,960 in the four-hop network, a reduction of twice the capacity compared to the previous case. The relationship developed in this section can be used to size the throughput (or the traffic carrying capacity) of a network serving multi-hop traffic with a specified maximum jitter bound and transmission resources limited to C bps.

6.4 Conclusions

This chapter has derived a quantitative relationship among the number of hops and end-to-end jitter in a packet switched network. Since jitter is an important factor in determining the Quality of Service of VoIP, the relationships derived can be used in sizing the resources of the network as the number of hops is increased. Alternatively, if the network resources were fixed, the relationships developed can be used to size the traffic bearing capacity of the network to serve VoIP traffic while keeping the end-to-end jitter limited to a pre-defined bound.

Chapter 7
Cost and Quality in Packet Switched Networks

Abstract The notion of quality is an important factor in the design of any complex communication system. Since communication systems are increasingly dependent on packet switching as the underlying transport mechanism, this chapter explores the impact of levels of quality that a packet switched system can provide on cost. Several interesting results are presented that result in the notion of an optimum level of load that a packet switched system can handle given a defined upper bound on latency that a specified percentage of packets would suffer. The results presented offer a fresh insight into the relationship between cost and quality and they can be used to optimize the throughput of a packet switched system under specified levels of quality. The contents of this chapter have been published in [103] and are reproduced with permission.

7.1 Introduction

The Internet is a prime example of a common user network that uses packet switching as its transport fabric. The growing reach of the Internet over the past decade together with the promise of lower costs to the customer has led to, for example, the rapid emergence of Voice over IP [104]. VoIP, being a real-time service, requires that qualities of service (QoS) parameters, such as delay, jitter or packet loss are tightly controlled. The relationship among these parameters can be developed using contemporary analytical techniques. This chapter shows that a packet switched network with a known level of transmission resources can handle an optimum amount of traffic for a specified level of QoS. This chapter develops that relationship. VoIP is an emerging service over the Internet. For most of the analysis and examples in this chapter, we use VoIP as the underlying application over a packet switching transport fabric.

P. K. Verma and L. Wang, *Voice over IP Networks*,
Lecture Notes in Electrical Engineering, 71, DOI: 10.1007/978-3-642-14330-4_7,
© Springer-Verlag Berlin Heidelberg 2011

The legacy circuit-switched network allocates dedicated 64 Kbits/s bandwidth to each voice call resulting in virtually no delay due to queuing and no jitter. Usually, PSTN provides a high level of voice quality, called toll quality [105]. VoIP packets in packet-switched networks undergo varying amounts of delay at each transit node. Delay in VoIP networks comes from three sources: processing, queuing and propagation. Therefore, to meet adequately the performance requirements of real-time application, delay becomes a significant parameter in VoIP networks. International Telecommunications Union—Telephony Recommendation G.114 [31] provides one-way transmission delay specifications for voice. Just as in the rest of the book, the delay addressed in this chapter is the *variable queuing delay*, the cumulative value of the delay each voice packet has to suffer at each router in the path of a VoIP connection.

We note that a system with bounded upper delay necessarily controls both the delay as well as the jitter. Taken from the service provider's perspective, a larger allowable range of delay variation provides more flexibility in terms of resource provisioning. The interplay between the needs of the user in terms of the QoS and the requirement it places on the service provider in terms of its impact on transmission resources to be provisioned is the primary issue addressed in this chapter.

A reasonable way to characterize the QoS in VoIP networks [106, 72] is by limiting the end-to-end queuing delay to an acceptable upper bound. We adopt this approach in this chapter. Packets that exceed the upper bound of delay (termed the threshold delay) are not counted as constituting effective throughput. The customer only pays for the effective throughput constituted by those packets that are within the delay threshold that characterizes the QoS received [107, 108]. As expected, VoIP traffic suffers higher queuing delay with increasing utilization of the link [79]. However, a higher utilization will also result in higher effective throughput under our construct. Increasing the QoS is thus tantamount to lowering the throughput that the packet switched service and, accordingly, a lower chargeable throughput. We explore the possibility of an optimum utilization of the link that results in the maximum throughput while at the same time keeping the QoS (as measured through the threshold delay), within the defined bound. Furthermore, this chapter also explores the impact of the number of hops on the end-to-end queuing delay of a voice packet. The resulting relationship leads to several interesting results that can be used to price Voice over IP services offered over a multi-hop network in a way that maximizes the utilization of the network (thus maximizing its throughput) while, at the same time, keeping the end-to-end queuing delay within a defined upper bound. We use the results to price appropriately VoIP services at equitable levels that are consistent with the resources consumed in order to achieve the contracted QoS.

The rest of the chapter is organized as follows: Sect. 7.2 presents a QoS based pricing model; Sect. 7.3 presents a mathematical model of VoIP networks; Sect. 7.4 presents the analytical results of pricing for the single-hop VoIP networks; Sects. 7.4 and 7.5 analyze the two-hop and multi-hop networks, respectively, as well as the pricing schemes; Sect. 7.6 analyzes one-node, two-node and multiple-node respectively; Sect. 7.7 presents the conclusion of this chapter.

7.2 A QoS Based Pricing Model

An important issue in designing pricing policies for today's networks is to balance the trade-off between traffic engineering and economic efficiency [16, 17]. A recent work [18] has addressed the impact of multiple hops (or switches between the ingress and egress switching nodes) on the grade of service offered by a circuit switched telephone network. The analytical results presented in that paper have led to proposing a new pricing scheme based on the cost of lost opportunity versus the cost of consumption of resources, which is the contemporary practice. We adopt a similar (but not identical) approach in this chapter in the context of packet switching. Accordingly, the grade of service is replaced by the threshold delay that is an appropriate measure for perceived QoS in a packet switched network. In a circuit switched network, an incomplete call is lost and does not generate any revenue. In packet switched networks, there are no calls that are lost as such; however, some of the packets may suffer delays above the acceptable delay bound and are, similarly, not considered to provide effective throughput. Just as a caller in a circuit switched network does not pay for an incomplete call, the VoIP caller over a packet switched network in our construct does not pay for packets that suffer an unacceptable level of delay.

7.3 Mathematical Model of a VoIP Network

A typical voice packet would pass through several nodes before arriving at the destination. We consider the voice traffic served by a single server as well as multiple servers. A server effectively constitutes a node of the network. Two or more servers are connected by transmission lines.

Consider a LAN shown in Fig. 4.2 with a VoIP (e.g., a SIP) server that functions as a VoIP network. We model the VoIP packets arriving at the SIP server as M/M/1 traffic [76]. A two-node tandem network is shown in Fig. 4.4. As in the literature [62], we analyze the QoS under the assumption that voice packets continue to follow the Poisson discipline at each intermediate node. Since our major interest in this chapter is to understand the impact of multiple nodes on QoSs, without loss of generality, we can assume the server capacity and the arriving traffic at each node to be identical.

We assume that all traffic served by the first node forms the incident traffic for the second node, *even if it was delayed beyond the threshold t.* In other words, the policy of discarding traffic with a delay higher than t is executed by the exit node.

7.4 Analysis of the Single-Hop VoIP Network

This section considers the impact of bounding delays on the capacity needed and the resulting throughput of the single-hop VoIP network.

7.4.1 Threshold Delay, Resource Consumption and Throughput

For an M/M/1 system, the probability that a voice packet suffers a delay less than t is given by [58, 81],

$$p_1 = P\{W_1 \leq t\} = 1 - \rho e^{-\mu C_1(1-\rho)t}. \tag{7.1}$$

Therefore, the throughput of the single-hop VoIP network where all packets that undergo a queuing delay not exceeding the threshold delay can be expressed as:

$$\gamma_1 = \lambda p_1, \tag{7.2}$$

where λ is the arrival rate of packets.

The normalized throughput, i.e., throughput expressed as a function of the incident traffic can be given as:

$$p_1 = \frac{\gamma_1}{\lambda}. \tag{7.3}$$

Using Eq. 7.1, Fig. 7.1 depicts the normalized throughput as a function of the threshold delay t with several values of the utilization factor $\rho(=\lambda/\mu C_1)$ as a parameter. The other parameters used are shown in the figure, $\lambda = 2,000$ and

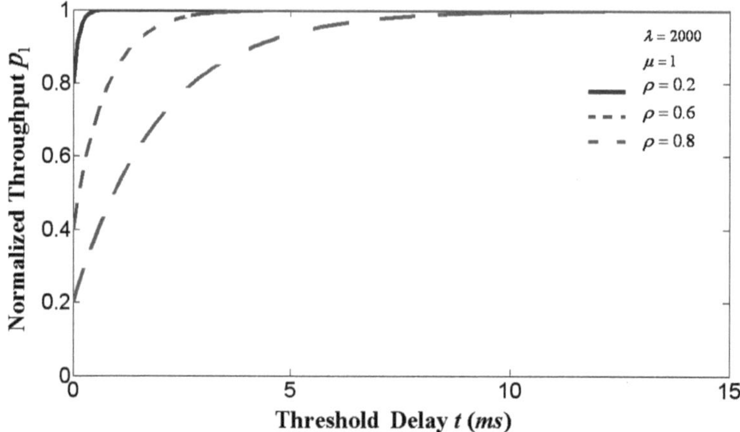

Fig. 7.1 Normalized throughput of single-hop network

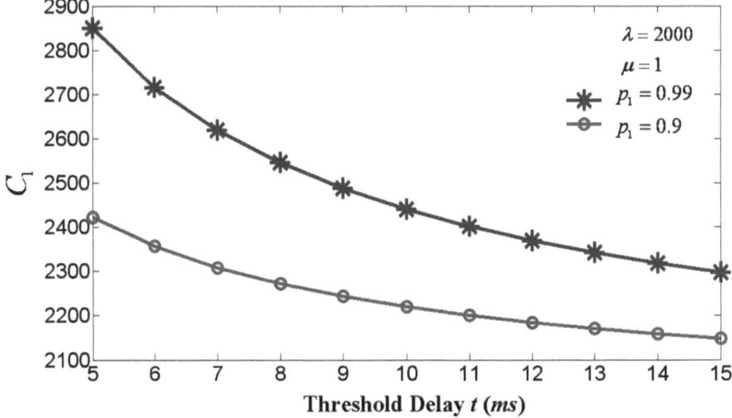

Fig. 7.2 Capacity as a function of threshold delay

$\mu = 1$, which are applied through the whole chapter. It can be seen that, given the same traffic intensity ρ, the normalized throughput increases as the threshold delay increases, while given the same threshold delay t, the normalized throughput increases as the traffic intensity decreases. In other words, the normalized throughput of light traffic load is higher than that of the heavy traffic load. In general, the customer would like a lower value of the threshold delay t, while the service provider's profit will increase as t increases because it would result in a higher throughput, which is paid for by the customer.

Equation 7.1 can also be used numerically to solve for the capacity C_1 needed for varying threshold delay t, with the normalized throughput p_1 used as a parameter, given values of λ and μ as shown in Fig. 7.2. It can be seen, as expected, that the capacity needed reduces as the threshold delay increases. In other words, a higher value of threshold will require a lower capacity. From Fig. 7.2, we can see that for a threshold delay of 8 ms, a 99% throughput is obtained for a specific value of $C_1 = 2,550$ bps. On the other hand, using a link capacity of 2,280 bps, only 90% packets are received with the same threshold delay.

From Fig. 7.1, it can be seen that the *relative gain* in normalized throughput reduces as the threshold delay is increased, even as the normalized throughput monotonically increases. It would be interesting to determine if the resulting throughput per unit of bandwidth used shows a point of inflexion.

From Eqs. 7.1 and 7.2, we have,

$$\frac{\gamma_1}{C_1} = \frac{\lambda}{C_1}\left[1 - \frac{\lambda}{\mu C_1}e^{-(\mu C_1 - \lambda)t}\right]. \tag{7.4}$$

A point of inflexion can be determined by putting

$$\frac{d(\gamma_1/C_1)}{dC_1} = 0 \tag{7.5}$$

which results in:

$$\mu C_{1opt} e^{(\mu C_{1opt} - \lambda)t} = \lambda\left(2 + \mu C_{1opt} t\right), \tag{7.6}$$

where C_{1opt} is the optimum capacity needed for the system to reach the maximum throughput per unit of bandwidth. Equation 7.4 shows a maximum value because

$$\frac{d^2(\gamma_1/C_1)}{dC_1^2} = \frac{2\lambda}{C_1^3}\left\{1 - \frac{\lambda}{2\mu C_1}[(\mu C_1 t + 2)^2 + 2]e^{-(\mu C_1 - \lambda)t}\right\} \tag{7.7}$$

is negative.

Using Eq. 7.4, γ_1/C_1 is plotted as a function of the bandwidth C_1 with several threshold delays used as a parameter, as shown in Fig. 7.3. From the service provider's perspective, the capacity C_1 is best utilized when the maximum γ_1/C_1 is at its peak. Thus, given a threshold delay t, the capacity C_{1opt} is the optimum capacity that the network requires. At this capacity, the user's requirement of delay is met while the throughput delivered per unit of bandwidth (or resource consumption) is maximized. From Fig. 7.3, we also notice that the maximized throughput per unit of bandwidth used increases as the threshold delay increases. In other words, a higher threshold delay results in a better utilization of the bandwidth.

We can evaluate the normalized throughput corresponding to the optimized capacity as follows. This will allow us to compute the throughput for which the customer will be actually charged.

From Eqs. 7.6 and 7.1 we get:

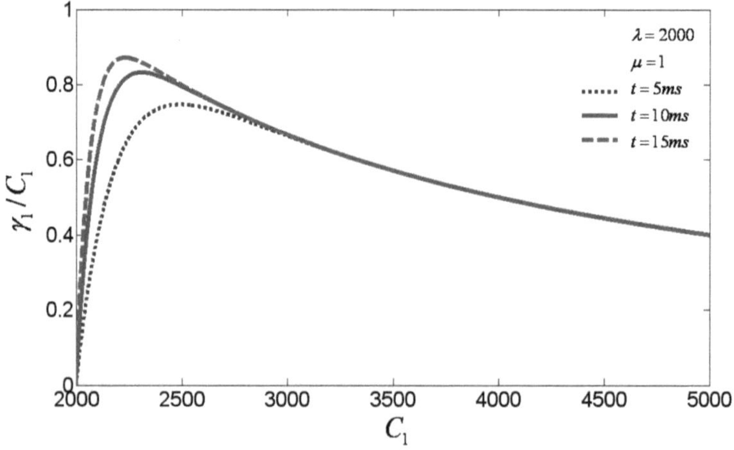

Fig. 7.3 Throughput/capacity as a function of capacity

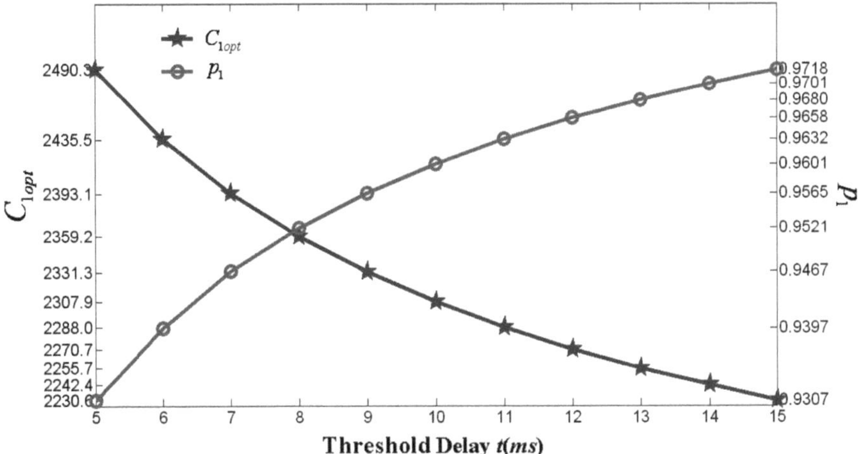

Fig. 7.4 Optimized capacity and normalized throughput as functions of threshold delay

$$p_{1opt} = 1 - \frac{1}{2 + \mu C_{1opt}t}, \tag{7.8}$$

where p_{1opt} is the normalized throughput corresponding to the optimum utilization of the bandwidth. Figure 7.4 plots the optimum capacity along with the normalized throughput as a function of the threshold delay. It can be easily observed that, as the threshold delay increases, the optimized capacity reduces while the normalized throughput increases.

It would be interesting to observe the impact of unit capacity increase on the corresponding reduction in the threshold delay t. Suppose the capacity necessary for supporting the threshold delay t_0 is C_0. If a specific customer were to subscribe for a higher QoS typified by a lower threshold delay t, where t is less than $t_0(t < t_0)$, then in order to achieve the same throughput, the new capacity C_1 needed can be obtained from Eq. 7.2 as follows:

$$\gamma_1 = \lambda\left[1 - \frac{\lambda}{\mu C_1}e^{-(\mu C_1 - \lambda)t}\right] = \text{const.} \tag{7.9}$$

Assuming λ and μ are fixed, the relation between the single-hop capacity C_1 and the threshold delay t can be computed by differentiation over t on both sides of Eq. 7.9,

$$\left\{\frac{\lambda^2}{\mu C_1^2}\frac{dC_1}{dt} + \frac{\lambda^2}{\mu C_1}\left[\mu t\frac{dC_1}{dt} + (\mu C_1 - \lambda)\right]\right\}e^{-(\mu C_1 - \lambda)t} = 0. \tag{7.10}$$

With simplification, we have,

$$\frac{dC_1}{dt} = -\frac{(\mu C_1 - \lambda)C_1}{1 + \mu C_1 t}. \tag{7.11}$$

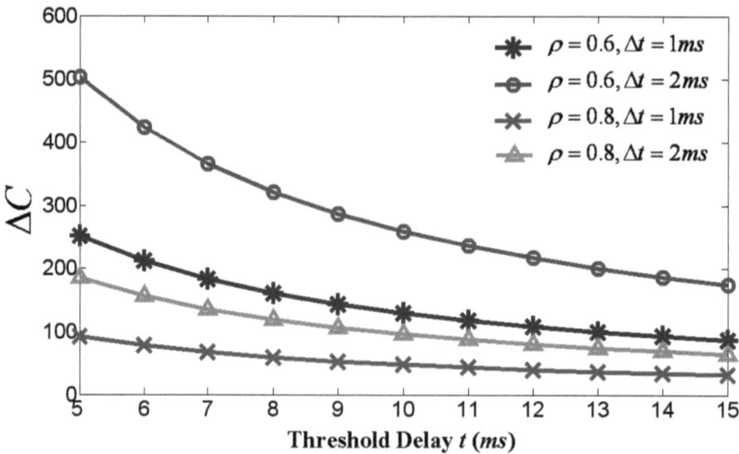

Fig. 7.5 ΔC as a function of threshold delay

The negative value illustrates the inverse relationship between an additional unit of capacity and the threshold delay. Using Eq. 7.11, Fig. 7.5 shows the impact of threshold delay t on ΔC, with ρ and Δt as parameters. It can be seen that for the single-hop network, a smaller ΔC is required in order to reduce the threshold delay by Δt as t increases to maintain the same throughput. We also observe that given t and Δt, a lighter traffic load (smaller ρ) results in larger capacity increase ΔC compared to when the network carries a heavy traffic load.

7.4.2 Pricing for Single-Hop Network

As we have seen already, from Eq. 7.6, given the characteristics of the incident traffic in terms of λ and μ, and a given requirement of the QoS in terms of the threshold delay t, the capacity needed to fulfill that requirement can be numerically evaluated.

In our VoIP network model, the discarded traffic will not generate any revenue from the customer. The service provider can only charge the customer of the served traffic [109]. The amount of served traffic λp_1 will determine the revenue for the service provider.

In the previous analysis, the single-hop VoIP network can function most efficiently when the capacity is best utilized [110]. In our pricing model for providing services with different QoS in the single-hop VoIP network, we treat the used capacity as the underlying cost and the throughput as the revenue. For the service provider, C_{1opt} represents the cost of providing service. The throughput achieved under this condition is λp_{1opt}. The cost of unit service (assuming bandwidth equals cost) is thus $C_{1opt}/\lambda p_{1opt}$. A customer achieving the throughput γ will thus be charged an amount equal to $(C_{1opt}/\lambda p_{1opt}) \times \gamma$.

If the service provider were to choose another threshold delay t', the necessary bandwidth C'_{1opt} as well as the corresponding p'_{1opt} can be similarly calculated. Then the unit price for throughput changes to $C'_{1opt}/\lambda p'_{1opt}$, and an individual customer realizing the throughput γ will be charged an amount equal to $\left(C'_{1opt}/\lambda p'_{1opt}\right) \times \gamma$.

We use the parameters presented in Fig. 7.4 as a specific example. We have $\lambda = 2000$, $\mu = 1$ and $t = 5$ ms. We compute $C_{1opt} = 2490.3$ and $p_{1opt} = 0.9307$. Therefore, the unit price can be obtained as $C_{1opt}/\lambda p_{1opt} = 2490.3/(2000 \times 0.9307) = \$1.34/\text{bps}$. If we increased the threshold delay to $t' = 6$ ms, C_{1opt} decreases to $C'_{1opt} = 2435.5$, and the normalized throughput increases to $p'_{1opt} = 0.9397$. The new unit price can be calculated the same way as before $C'_{1opt}/\lambda p'_{1opt} = \$1.3/\text{bps}$, which is less than the unit price charged for the first service with 1 ms less delay threshold.

In view of the results presented in Eqs. 7.6–7.8 and the above example, we can state the pricing strategy for a single-hop VoIP network as follows.

For the traffic parameters λ, μ and the QoS defined by the threshold delay t associated with the user's needs, compute the optimum channel capacity C_{1opt} from Eq. 7.6. The nominal unit price for the network is computed as,

$$\text{Unit price} = \frac{C_{1opt}}{\lambda p_{1opt}}. \tag{7.12}$$

For a particular customer generating traffic at the rate of $\lambda_1(\lambda_1 < \lambda)$, the price can be determined as:

$$\frac{C_{1opt}}{\gamma} \times \gamma_1 = \frac{C_{1opt}}{\lambda p_{1opt}} \times \lambda_1 p_{1opt} = C_{1opt}\frac{\lambda_1}{\lambda} \tag{7.13}$$

which is proportional to the price of the total incident traffic.

7.5 Analysis of Two-Hop VoIP Network

7.5.1 Comparisons of Two-Hop and Single-Hop Traffic Performance

We now consider the throughput of a VoIP network with two hops between the sending and receiving nodes. The arrival process of the traffic incident on the second node is assumed to be Poisson as well [78]. As described in Sect. 7.2, the last or the exit node drops the packets that have suffered delay higher than the threshold delay t. The probability density function (pdf) of the waiting time $f_{W_2}(t)$ in the two-hop network can be computed by convolving the corresponding pdfs of

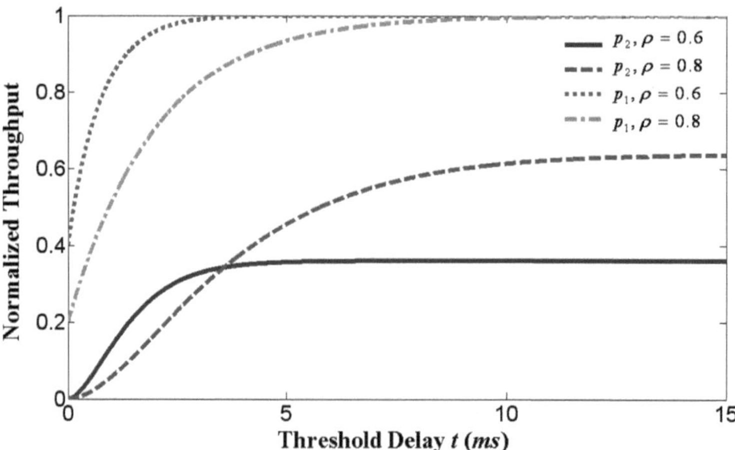

Fig. 7.6 Normalized throughput for single-hop and two-hop systems

the waiting time at each node $f_{W_1}(t)$ [82]. Therefore, the waiting time distribution can be given as shown in (4.13),

$$p_2 = P(W_2 \le t) = \rho^2 \left\{ 1 - e^{-\mu C_2(1-\rho)t}[1 + \mu C_2(1-\rho)t] \right\}. \tag{7.14}$$

The throughput can now be given as

$$\gamma_2 = \lambda p_2. \tag{7.15}$$

Therefore, the normalized throughput for a two-hop system can be expressed as

$$p_2 = \frac{\gamma_2}{\lambda}. \tag{7.16}$$

We have plotted Eqs. 7.1 and 7.14 in Fig. 7.6 for specified values of p_2 and ρ. The same figure also reproduces corresponding curves for the single node network. We observe the following two characteristics: the normalized throughput for both the single-hop and two-hop networks increases as a function of t; however, in contrast to the single-hop network, the two-hop network normalized throughput never reaches 100%. This can also be observed from Eq. 7.14, where p_2 will always remain lower than ρ^2. A two-hop network can thus never achieve 100% throughput if we measured throughput as served traffic that suffers delay below a specified upper bound, t.

Another interesting observation from Fig. 7.6 is that, for the two-hop network, a lower utilization factor results, initially, in a higher normalized throughput; however, the higher utilization factor leads to a higher normalized throughput beyond a certain value of the threshold delay. The point of intersection where the higher utilization starts delivering a higher normalized throughput can be numerically evaluated from Eq. 7.14.

We further note that if the two-hop network were to meet the latency requirement, since it has two links, its capacity cost would be $2C_1$ because of the two hops involved. (We assume that the cost of capacity is independent of distance and is directly proportional to the number of hops.) However, since the carried two-hop traffic is always lower than the carried single-hop traffic with identical capacity in each hop, as from Eq. 7.14 and as illustrated in Fig. 7.6, it follows that the two-hop network should be priced more than twice the single-hop price. Mathematically, from (7.1), (7.2), (7.14) and (7.15), we have,

$$\frac{\gamma_2}{\gamma_1} = \frac{\lambda\left(\frac{\lambda}{\mu C_1}\right)^2\left\{1 - e^{-(\mu C_1 - \lambda)t}[1 + (\mu C_1 - \lambda)t]\right\}}{\lambda\left[1 - \frac{\lambda}{\mu C_1}e^{-(\mu C_1 - \lambda)t}\right]} \tag{7.17}$$

which is always less than 1.

The actual price for VoIP services that involve two links instead of one will be determined by the enhanced capacity needed in each of the two links that would result in identical threshold delay applicable to the single-hop network. Suppose this capacity were C_2 for each of the two links in the two-hop network. Then, for the throughput to be identical, we must have

$$\gamma_2 = \gamma_1 \tag{7.18}$$

or

$$\left(\frac{\lambda}{\mu C_2}\right)^2\left\{1 - e^{-(\mu C_2 - \lambda)t}[1 + (\mu C_2 - \lambda)t]\right\} = 1 - \frac{\lambda}{\mu C_1}e^{-(\mu C_1 - \lambda)t}. \tag{7.19}$$

From (7.19), the value of C_2 can be numerically evaluated.

Figure 7.7 plots C_2/C_1 as a function of the threshold delay t. We observe that the relative capacity per link for the two-hop network decreases as the threshold delay increases. Furthermore, given t, the required C_2/C_1 increases as the normalized throughput increases.

The maximum normalized throughput p_{2opt} can be obtained by putting the first derivative of p_2 equal to zero. In other words, p_2 will be maximized for the specific C_2 such that

$$\frac{dp_2}{dC_2} = 0, \quad \text{where } \frac{d^2p_2}{dC_2^2} < 0 \tag{7.20}$$

(The inequality can easily be shown to be true.)

(7.20) results in the following transcendental equation:

$$2e^{(\mu C_{2opt} - \lambda)t} = 2 + \left(2 + \mu C_{2opt}t\right)\left(\mu C_{2opt} - \lambda\right)t. \tag{7.21}$$

C_{2opt} is the optimum capacity, at which the normalized throughput p_2 of a two-hop VoIP Network is maximized. It is characterized by a fixed arriving traffic λ. All packets incurring a queuing delay higher than t are discarded.

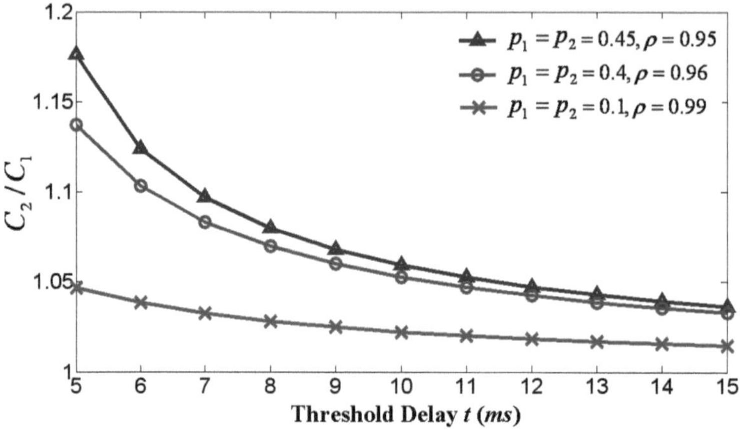

Fig. 7.7 Relative capacity of tandem network as a function of threshold delay t

7.5.2 Pricing for Two-Hop Network

Similar to the pricing model for the single-hop network, for the two-hop VoIP network with design parameters λ, μ and t, the unit price for the two-hop traffic can be observed to be $\$2C_{2opt}/p_{2opt}$/bps, where p_{2opt} is the normalized throughput using C_{2opt} as the capacity of each of the two links. With another chosen threshold delay t', the necessary bandwidth C'_{2opt} per link and maximum normalized throughput p'_{2opt} can also be calculated from Eqs. 7.21 and 7.14. In that case a different unit price $\$2C'_{2opt}/\lambda p'_{2opt}$ will be charged to the customer subscribing to this service. The total price will be $(2C'_{2opt}/\lambda p'_{2opt}) \times \gamma$ based on the served traffic γ.

As an example, given $\lambda = 2,000$, $\mu = 1$ and threshold delay $t = 5$ ms, the numerical results are calculated as $C_{2opt} = 2649.9$ and $p_{2opt} = 0.4757$. Therefore, the unit price will be $2C_{2opt}/\lambda p_{2opt} = \$5.57$/bps. Recall from the example in Sect. 7.4 applicable to the single-hop traffic under the identical conditions that $C_{1opt} = 2490.3$ and the unit price $C_{1opt}/\lambda p_{1opt} = \$1.34$/bps. After comparison of $C_{2opt}/C_{1opt} = 1.064$ and $(2C_{2opt}/\lambda p_{2opt})/(C_{1opt}/\lambda p_{1opt}) = 4.1567$, we observe that the unit price for two-hop traffic is more than four times the unit price for single-hop traffic. The price increases because of two reasons: (1) Two links of identical capacity are needed, and (2) The bandwidth of each of the two links is higher because the required throughput for links of identical capacity is much lower in the case of two links in tandem. We consider another example when the threshold delay changes to $t' = 6$ ms. We have $C'_{2opt} = 2571.8$ and $p'_{2opt} = 0.5180$. The unit price $2C'_{2opt}/\lambda p'_{2opt} = \$4.9649$/bps, which is less than the capacity needed for the lower threshold delay of 5 ms. Compared to the single-hop traffic with the same threshold delay $t' = 6$ ms, $C'_{1opt} = 2435.5$ and $C'_{1opt}/\lambda p'_{1opt} = \$1.3$/bps, we have $C'_{2opt}/C'_{1opt} = 1.056$ and $(2C'_{2opt}/\lambda p'_{2opt})/(C'_{1opt}/\lambda p'_{1opt}) = 3.8192$. The price for a higher threshold delay thus shows an improvement compared to the price for a lower threshold delay case. The example also shows that both the ratios of the capacity per each link and the unit price of two-hop network relative to single-hop network decrease as the threshold delay increases. With the analysis and examples presented above, we can now state the pricing strategy for a two-hop VoIP network as follows.

For the design parameters, λ, μ and t associated with a two-hop network, we can compute the optimum channel capacity C_{2opt} from Eq. 7.21. The unit price for the network is now computed as,

$$\text{Unit price} = \frac{2C_{2opt}}{\lambda p_{2opt}}. \tag{7.22}$$

For a specific customer with incident traffic of $\lambda_2(\lambda_2 < \lambda)$, the price will be proportional to the price of the total incident traffic as

$$\$\frac{2C_{2opt}}{\lambda p_{2opt}} \times \lambda_2 p_{2opt} = \$2C_{2opt}\frac{\lambda_2}{\lambda}. \tag{7.23}$$

7.6 Analysis of Multi-Hop VoIP Network

For the multi-hop network, we assume that voice packets continue to follow the Poisson discipline at each intermediate node [80]. Under this assumption, p_h is computed by the convolution of probability density function (pdf) of waiting time associated with the first $(h-1)$-hop and the hth hop, as derived in (4.20). We get,

$$p_h = P(W_h \leq t) = \rho^h \left(1 - e^{-\mu C(1-\rho)t} \sum_{k=1}^{h} \frac{[\mu C(1-\rho)t]^{h-k}}{(h-k)!}\right). \tag{7.24}$$

As before, we can obtain the throughput of the multi-hop network as:

$$\gamma_h = \lambda p_h \tag{7.25}$$

As a function of p_1 with an identical threshold delay t, p_h can be expressed as:

$$p_h = \rho^h \left\{1 - \frac{1-p_1}{\rho} \sum_{k=1}^{h} \frac{\left[\ln\left(\frac{\rho}{1-p_1}\right)\right]^{h-k}}{(h-k)!}\right\}. \tag{7.26}$$

Using Eq. 7.26, Fig. 7.8 plots p_h as a function of the number of hops for given p and ρ. We can readily observe the sharp decline in the served traffic from one hop to two hops, and relatively smoother decrease from two hops to multiple hops.

Also, C_h/C_1 as the function of number of hops h is numerically evaluated from the Eqs. 7.1 and 7.24, with design parameters $\lambda = 2,000$ and $\mu = 1$. Figure 7.9

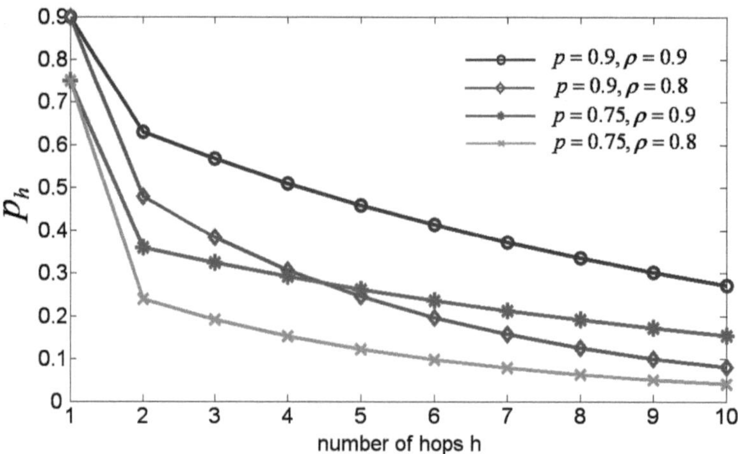

Fig. 7.8 p_h as a function of h with p and ρ as parameters

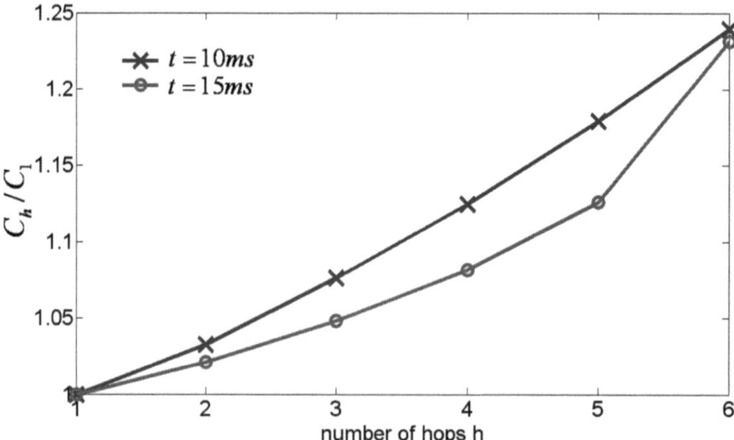

Fig. 7.9 Price of h-hop traffic

shows that, for example, for a given delay bound of 10 ms with four-hop traffic, the transmission resources needed at each hop is 1.08 times as much as that of the single-hop traffic. With the same $1.08C_1$ capacity of each link in a five-hop VoIP system, voice packets that have the same relative throughput will experience a delay threshold of 15 ms.

As discussed in Sect. 7.2, the customers are only charged for the served traffic. For multi-hop voice traffic, the pricing will depend on the number of hops as well as on the relative change in capacity of each hop compared to the single hop case with identical QoS. The procedure for pricing for multi-hop VoIP transport can now be stated as follows:

(1) Given the values of λ, μ and the threshold delay t, compute the optimized transmission resources C_{hopt} needed at each hop in an h-hop system;
(2) Compute the normalized throughput p_{hopt} in case of C_{hopt};
(3) The unit price will be obtained as $hC_{hopt}/\lambda p_{hopt}$;
(4) The price for the served traffic γ will be $(hC_{hopt}/\lambda p_{hopt}) \times \gamma$.

We note that the price levels derived above are applicable for a network that has optimized its transport resources so as to match the level of demand as specified through the incident traffic and the QoS as specified through the threshold delay.

7.7 Conclusion

This chapter has presented the impact of QoS demanded by a customer on the transport capacity of a VoIP network. We have developed analytical results that can lead to a determination of the optimum network capacity needed for prescribed levels of QoS. The customer's pricing is based on the served traffic that meets the QoS as specified through the threshold delay. Separate results are derived for single-, two- as well multi-hop networks. The deleterious impact of multiple hops on the resources needed to maintain the same QoS as the single-hop network can be quantitatively assessed as a result of the investigation reported in this chapter.

Chapter 8
Cumulative Impact of Inhomogeneous Channels on Risk

Abstract Chapter 4 developed a mechanism to evaluate the impact of more than one telecommunication channel in tandem on the distribution of delays. This chapter applies the mechanism developed in Chap. 4 in assessing the cumulative impact of risks constituted by a number of channels, where the risk associated with each channel can be quantified by a known distribution. More specifically, this chapter considers flows of containerized traffic in a cascade of channels with diverse risk characteristics. Each channel is characterized by a probability distribution function relating the probability of loss being less than a specified value to the magnitude of the loss. The cumulative impact of cascading channels is then evaluated as a closed form solution in terms of the characteristics of the constituent channels with dissimilar risk characteristics. The results presented in this analysis can be used to shape the risk characteristics of individual channels through, for example, additional investment in order to maximize the impact of such investments. The contents of this chapter have partially been published in [111] and are reproduced with permission.

8.1 Introduction

This chapter applies the mechanism developed in Chap. 4 in assessing the cumulative impact of risks constituted by a number of channels, where the risk associated with each channel can be quantified by a known distribution. The risk associated with a channel is in a mathematical sense equivalent to the delay experienced by a communication channel. The previous analysis has focused on computing the probability of communication channels in tandem having the cumulative delay bounded to a specified value. This chapter extends that analysis and applies it to a situation where one needs to evaluate the probability that the cumulative risk is bounded to a specified value [112, 113]. The specific example considered in this chapter is the flow of goods across international boundaries. This

P. K. Verma and L. Wang, *Voice over IP Networks*,
Lecture Notes in Electrical Engineering, 71, DOI: 10.1007/978-3-642-14330-4_8,

is actually typified by the flow of containerized cargo through a number of transportation channels, e.g., roads, ocean, or air. The flow of containerized traffic will, generally speaking, also encounter a variety of gates that include national boundaries, customs and other government-mandated checkpoints. Each of the modalities of transportation and the gates encountered by the container during its transit from the source to the destination presents a varying risk profile [114]. Each would influence the end-to-end risk characteristics of containerized traffic flow in complex ways.

This chapter develops a closed form solution relating the end-to-end risk behavior of containerized traffic flow in terms of the characteristics of each of the channels or gates. Understanding the end-to-end risk characteristics is important from the perspective of business because business needs to develop a predictable model for risk in order to insure sufficiently its cargo and factor the costs of such insurance in the pricing model [115]. From a national security perspective, each government or national security agency needs to weigh carefully the costs of improving safety and security against the predicted enhancement in attaining such security on an end-to-end basis. While the impact of an investment toward the improvement in the characteristics of a single channel or gate might be easily understood, this understanding is not sufficient in terms of evaluating the impact of that investment from an overall risk mitigation perspective when several channels in sequence are involved. Accordingly, it might not yield the best 'bang for the buck' invested. This chapter characterizes the end-to-end risk profile in terms of the characteristics of each of the constituent elements in order to maximize the impact of investment toward improving the end-to-end risk characteristics.

The conventional way of understanding risk is in terms of the expectation of loss, which is a product of the probability of loss and the average amount of loss. This information is insufficient if one were to assess the probability of loss remaining bounded within a predefined threshold. This idea is very similar to the concept developed in Chap. 4, where we showed that for telecommunication applications, the average delay is not a meaningful parameter. Accordingly, this chapter addresses the probability associated with specified bounds of risk rather than the value of the average risk associated with a number of transportation channels/gates in tandem. More specifically, it addresses the scenario that a cargo goes through when it traverses a number of channels that are inhomogeneous in their risk characteristics [116]. The risk characteristics of a single channel are first addressed followed by two channels in tandem. Then, the results are generalized to include a number of cascaded channels with inhomogeneous risk characteristics. The findings of the investigation are illustrated through a number of examples.

8.2 The Single Channel Model

The model defines the risk characteristics of each intermediate traffic channel and each gate encountered by the in-transit cargo individually. It is assumed that the risk characteristic of each element (whether a channel or a gate) is independent. The risk characteristics of each element are defined by a single variable λ which is exponentially distributed.

The probability density function of this variable is given by:

$$f(x) = \begin{cases} \lambda e^{-\lambda x}, & x \geq 0 \\ 0, & x < 0 \end{cases} \tag{8.1}$$

The cumulative distribution function is given by:

$$F(x) = \begin{cases} 1 - e^{-\lambda x}, & x \geq 0 \\ 0, & x < 0 \end{cases} \tag{8.2}$$

An interpretation of Eq. 8.2 would be that the probability of a loss of magnitude x or less is $F(x)$. The boundary conditions of Eq. 8.2 are verified by the fact that all losses are bounded by infinity and zero. Using the properties of the exponential distribution, the mean loss of a channel $= 1/\lambda$. Figure 8.1 shows the distribution function for three different values of $\lambda = 0.5$, 1, and 1.5. The curve with a higher value of the mean loss (lambda $= 0.5$) rises more slowly as expected. Figure 8.1 shows the distribution function for three different values of $\lambda = 0.5$, 1, and 1.5.

8.3 The Two-channel Model

A two-channel model is considered, say, consisting of road and air transport channels. The probability density function of the two channels is defined as:

$$f_1(x) = \lambda_1 e^{-\lambda_1 x} \tag{8.3}$$

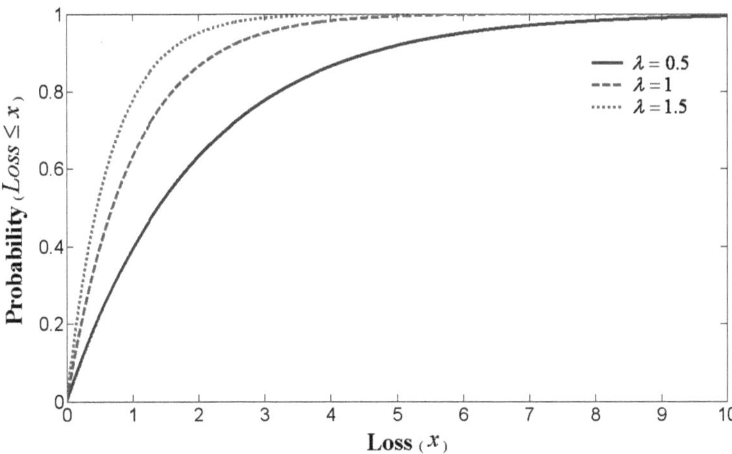

Fig. 8.1 Loss Characteristics of a single channel

$$f_2(x) = \lambda_2 e^{-\lambda_2 x} \tag{8.4}$$

The combined probability density function $f(x)$ of both the channels can be evaluated by convolving the 2 constituent probability density functions $f_1(x)$ and $f_2(x)$ [117, 118]. We have,

$$f(x) = f_1(x) \otimes f_2(x) = \int_{-\infty}^{\infty} f_1(\xi) f_2(x - \xi) \mathrm{d}\xi$$

$$= \int_{-\infty}^{\infty} \lambda_1 e^{-\lambda_1 \xi} \lambda_2 e^{-\lambda_2 (x-\xi)} \mathrm{d}\xi = \lambda_1 \lambda_2 e^{-\lambda_2 x} \int_{0}^{x} e^{(\lambda_2 - \lambda_1)\xi} \mathrm{d}\xi$$

$$= \frac{\lambda_1 \lambda_2}{\lambda_2 - \lambda_1} e^{-\lambda_2 x} (e^{(\lambda_2 - \lambda_1)x} - 1)$$

$$= \frac{\lambda_1 \lambda_2}{\lambda_2 - \lambda_1} (e^{-\lambda_1 x} - e^{-\lambda_2 x})$$

$$\tag{8.5}$$

For $\lambda_1 = \lambda_2 = \lambda$,

$$f(x) = \int_{-\infty}^{\infty} \lambda e^{-\lambda \xi} \lambda e^{-\lambda (x-\xi)} \mathrm{d}\xi = \int_{0}^{x} \lambda^2 e^{-\lambda x} \mathrm{d}\xi = \lambda^2 x e^{-\lambda x} \tag{8.6}$$

After some simplification, Eq. 8.5 can be expressed as,

$$f(x) = \begin{cases} \frac{\lambda_1 \lambda_2}{\lambda_2 - \lambda_1}(e^{-\lambda_1 x} - e^{-\lambda_2 x}), & (\lambda_1 \neq \lambda_2) \\ \lambda^2 x e^{-\lambda x}, & (\lambda_1 = \lambda_2 = \lambda) \end{cases} \tag{8.7}$$

$$F(x) = \begin{cases} 1 - \frac{1}{\lambda_2 - \lambda_1}(\lambda_2 e^{-\lambda_1 x} - \lambda_1 e^{-\lambda_2 x}), & (\lambda_1 \neq \lambda_2) \\ 1 - (1 + \lambda x)e^{-\lambda x}, & (\lambda_1 = \lambda_2 = \lambda) \end{cases} \tag{8.8}$$

From Eq. 8.8, it is noticed that λ_1 and λ_2 are interchangeable. Therefore, the sequence of the channels does not affect the end-to-end loss. In other words, the end-to-end loss characteristic is determined entirely by the loss characteristics of each channel and is independent of their sequence.

Examples Figure 8.2 presents the loss characteristics of four different situations, each with two channels and the following loss characteristics.

Case 1: $\lambda_1 = 0.75$, $\lambda_2 = 1.25$
Case 2: $\lambda_1 = 0.5$, $\lambda_2 = 1.5$
Case 3: $\lambda_1 = 0.25$, $\lambda_2 = 1.75$
Case 4: $\lambda_1 = 1$, $\lambda_2 = 1$

Note that in each case the sum of λ_1 and λ_2 has been kept at a constant value, namely, $\lambda_1 + \lambda_2 = 2$.

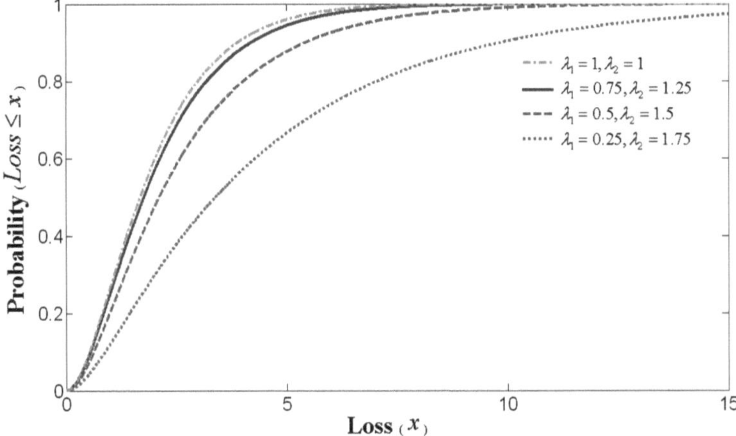

Fig. 8.2 Loss Characteristics of two channels in tandem

It is evident from Fig. 8.2 that closer values of λ_1 and λ_2 result in curves that represent better end-to-end loss characteristics, i.e., the curves rise faster than those where λ_1 and λ_2 widely vary.

It is noted in passing that the mean value of loss of two channels in tandem is equal to $\frac{1}{\lambda_1} + \frac{1}{\lambda_2}$. This sum will obviously increase as λ_1 and λ_2 diverge while their sum remains constant. In other words, the cumulative loss characteristics of two channels in tandem are consistent with the loss experienced by each of the two channels. Let

$$\lambda_1 + \lambda_2 = C(\text{a constant}) \tag{8.9}$$

It is intended to show that $\frac{1}{\lambda_1} + \frac{1}{\lambda_2}$ is minimized when $\lambda_1 = \lambda_2$. Let

$$\lambda_2 - \lambda_1 = \Delta \tag{8.10}$$

It is assumed, without loss of generality that $\lambda_2 > \lambda_1$, i.e. $\Delta > 0$. From Eq. 8.9 and 8.23, we have $\lambda_2 = \frac{C+\Delta}{2}$ and $\lambda_1 = \frac{C-\Delta}{2}$.

After some algebraic simplification, it is

$$\frac{1}{\lambda_1} + \frac{1}{\lambda_2} = \frac{4C}{C^2 - \Delta^2} \tag{8.11}$$

For C and $\Delta > 0$, it can be easily shown that (8.11) is minimized when $\lambda_1 = \lambda_2$.

Figure 8.3 plots a number of curves for two channels in random where the sum $\frac{1}{\lambda_1} + \frac{1}{\lambda_2}$ is kept a constant while varying the individual λ_1 and λ_2. It can be observed from Fig. 8.3 that, as long as the mean end-to-end loss experienced is kept to be a constant, the variance in their cumulative loss characteristics is moderate.

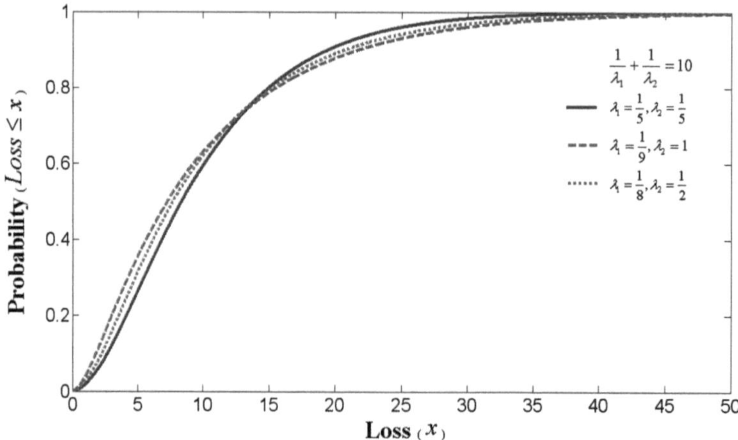

Fig. 8.3 Loss Characteristics of two channels with the same end-to-end mean loss

We can now ask ourselves the following question: Given identical end-to-end mean loss, is it preferable to have a single channel or two channels in tandem? A surprisingly elegant result is presented in Theorem 1.

Theorem 1 Compared to two identical channels in tandem, each having a loss parameter equal to λ, a single-channel with the parameter $\frac{\lambda}{2}$ has superior cumulative loss characteristics up to a mean loss value that can be numerically evaluated. Beyond this point, the two-channel model with identical mean loss is superior.

Proof It is noted that the mean loss value for each model is identical since $\frac{1}{\lambda} + \frac{1}{\lambda} = \frac{1}{\frac{\lambda}{2}}$. We also have, for its single-channel model,

$$F_1(x) = 1 - e^{-\frac{\lambda}{2}x} \tag{8.12}$$

And for the two-channel model

$$F_2(x) = 1 - (1 + \lambda x)e^{-\lambda x} \tag{8.13}$$

For very small values of x, we have

$$F_1(x) \approx \frac{\lambda}{2}x - \frac{\lambda^2 x^2}{8} \tag{8.14}$$

$$F_2(x) \approx \frac{\lambda^2 x^2}{2} \tag{8.15}$$

And therefore, for such values of x, $F_1(x) > F_2(x)$

The two loss curves intersect when $F_1(x) = F_2(x)$, i.e., when from Eqs. 8.12 and 8.13

$1 - e^{-\frac{\lambda}{2}x} = 1 - (1 + \lambda x)e^{-\lambda x}$, or when

$$(1 + \lambda x)e^{-\frac{\lambda}{2}x} - 1 = 0 \tag{8.16}$$

Equation 8.16 can be numerically evaluated.
For large values of x, we have,

$$F_2(x) - F_1(x) = e^{-\frac{\lambda}{2}x} - (1 + \lambda x)e^{-\lambda x} \tag{8.17}$$

The ratio of the two terms on the R.H.S. of Eq. 8.17 can be written as,

$$\frac{e^{-\frac{\lambda}{2}x}}{(1 + \lambda x)e^{-\lambda x}} = \frac{e^{\frac{\lambda}{2}x}}{1 + \lambda x} \tag{8.18}$$

The problem then becomes to compare the relative values of $e^{\frac{\lambda}{2}x}$ and $1 + \lambda x$. For a given value of λ, it is further noted that since the *slope* of an exponential function, $\frac{\lambda}{2}e^{\frac{\lambda}{2}x}$ increases with x, while the slope of a straight line is constant λ, there cannot be more than two points of intersection between a straight line and an exponential curve. The two channel model thus has better loss characteristics than the single channel model at higher values of loss.

8.4 The n-Channel Model

For n-channel model, there are two cases. One supposes that each channel has the same λ, then

$$f_n(x) = \frac{x^{n-1}}{(n-1)!}\lambda^n e^{-\lambda x} \tag{8.19}$$

$$F_n(x) = 1 - \sum_{i=0}^{n-1} \frac{(\lambda x)^i}{i!} e^{-\lambda x} \tag{8.20}$$

The other supposes λ, as

$$f_3(x) = \frac{\lambda_1 \lambda_2}{(\lambda_2 - \lambda_1)(\lambda_3 - \lambda_1)}e^{-\lambda_1 x} + \frac{\lambda_1 \lambda_2}{(\lambda_1 - \lambda_2)(\lambda_3 - \lambda_2)}e^{-\lambda_2 x}$$
$$+ \frac{\lambda_1 \lambda_2}{(\lambda_1 - \lambda_3)(\lambda_2 - \lambda_3)}e^{-\lambda_3 x} \tag{8.21}$$

$$f_n(x) = \sum_{i=1}^{n} \frac{\prod_{i=1}^{n-1} \lambda_i}{\prod_{j=1, j\neq i}^{n} (\lambda_j - \lambda_i)} e^{-\lambda_i x} \tag{8.22}$$

$$F_n(x) = 1 - \sum_{i=1}^{n} \frac{\prod_{i=1}^{n-1} \lambda_i}{\prod_{j=1, j\neq i}^{n} (\lambda_j - \lambda_i)} e^{-\lambda_i x} \tag{8.23}$$

8.5 Conclusion

This chapter has provided a closed form solution to the general problem of assessing the probability of bounding loss in a cascade of channels when the risk characteristics of each channel, modeled as an exponential loss model, are known. The results obtained can be utilized to shape the loss characteristics of an individual channel.

8.6 Further Extensions of the Technique

We have adopted a relatively straightforward distribution of the risk characteristics of the transportation channels in this chapter. It is well known that natural or man-made disasters do not always follow the statistical model typified by the exponential distribution. One model that has adequately represented natural calamities in the past is the lognormal distribution. The lognormal distribution is defined as the distribution of a random variable whose logarithm is normally distributed. It has been widely applied in real life for fitting data [119]. One specific example for the applicability of the lognormal distribution is the earthquake statistics [120].

The techniques described in this chapter can be easily extended to include any general distribution. It is unlikely, however, that a closed form solution for the distribution of cumulative risk can be obtained for distributions other than the exponential distribution. However, numerical techniques can be used to evaluate the cumulative risk characteristics of channels in tandem in all situations following the techniques proposed in this chapter.

Chapter 9
The Economics of VoIP Systems

Abstract So far, this book has been devoted to developing an understanding of the consumption of bandwidth resources related to VoIP systems. As important as this resource is, numerous other factors determine the price that the VoIP user would pay. This chapter reviews those factors and shows how they influence pricing. The first few sections of the chapter discuss the characteristics of information and the economics of information networks in general. The chapter then presents a model of the legacy network and identifies how the network is continuing to evolve to accommodate needs of VoIP service offerings. A path leading toward a potential integration of the legacy network and the Internet is indicated as well.

9.1 Introduction

Even under ideal conditions, pricing of communication services is a complex task. In a competitive marketplace, pricing and quality of service are the two primary mechanisms through which a service provider attracts customers as well as protects the financial integrity of the business. The telecommunication marketplace is no exception. Indeed, pricing of services in this market is especially complex as this market rapidly evolves from a monopoly to a competition-driven play. Additionally, specificities of the telecommunications market are not applicable to other commodity or specialty markets. Although not generally viewed as such, in some ways, telecommunication services do function as a commodity. For example, efficient communication systems can offset the use of steel in building additional railroad tracks by increasing the utilization of fewer tracks through the adoption of better signaling techniques. Similarly, the use of video conferencing can offset travel using airplanes or other modes of transportation. The question then arises as to why telecommunications services are viewed so differently as to engage the

P. K. Verma and L. Wang, *Voice over IP Networks*,
Lecture Notes in Electrical Engineering, 71, DOI: 10.1007/978-3-642-14330-4_9,

close attention of state and federal governments and the judicial processes, all over the world.

The specific characteristics of telecommunication services arise from two major factors. First, we note that telecommunications is the act of transporting information between and among entities. If we look at information as consumer goods, it has the very special property that it can be indefinitely replicated at very little cost. Further, it can be indefinitely stored with no noticeable degradation at little and sharply declining cost. Information and its communication or storage is thus very distinct from many other consumer goods and their transportation or storage. These differences arise because technology is so heavily embedded in telecommunications. Moreover, the cost of technology associated with information and its transportation continues to decline exponentially, thus making telecommunication systems continuing to offer a higher level of quality and user satisfaction. Unquestionably, this trend is likely to continue in the near future.

Telecommunication networks are capital intensive in the sense that establishing a network requires a large amount of capital in order to serve the very first customer. Capital is expended, for example, in acquiring the right of way to lay fiber optic or other transmission media, in the labor and material costs associated with laying the cable, in the cost associated with switching or routing equipment, or in the cost incurred in buying blocks of spectrum for offering wireless services. Since information is the commodity that is transported over the network, the characteristics of information largely define the economic parameters associated with providing services offered by the network. We review some of the characteristics of information and communication networks in the following sections.

9.2 Economy of Scale of Communication Networks

The value of a network is known to increase as the square of the number of end points or clients it serves. One way to visualize this is the fact that a network of n end points can have up to $n(n-1)/2$ bilateral associations. A network with a single customer cannot enable any communication; with two customers, the network could provide a maximum of just one communication path. Four customers on a network can have up to six-one-on-one associations, and so on. Because of this, there is a built in incentive for customers to be served by networks that provide the most number of end points.

Robert Metcalfe, inventor of the Ethernet, articulated this 'law', widely known as the Metcalfe's Law to quantify the value of a network as a function of the number of its end points. It should be emphasized that Metcalfe's Law is not an immutable physical law but merely an attempt to explain a phenomenon that was observed during information networking's boom years. During the late eighties through the early nineties, information network related businesses grew at a dizzying speed. This growth was driven by the widespread availability of Personal Computers, the emerging Internet, and the browser technology, all at the same time. However, the law provided an easy explanation for growth stating that the value of information networks will continue to increase quadratically as the number of subscribers. This realization on part of the investment community gave rise to an enormous appetite for investment to build and grow networks. The underlying assumption was that the increase in value—and therefore profitability—will continue to increase unabated indefinitely. This has not been the case, however. The overinvestment in information networking businesses has resulted in increasing losses and has penalized several corporations.

More recently, the aggressive Metcalfe's law has been made more sober by relating the value of a network to $n \log (n)$ instead of n squared [121]. The fact, however, does remain that a network's value and, therefore, profitability is related to its size. In other words, there will always be an incentive for a network operator to grow the size of its network.

Just as a network of large reach, in terms of the number of endpoints it serves, is important to the customer, the size of the network is also important to the network provider from another perspective. This is because of the economy of scale associated with networking equipment. The network provider would thus like to capture as many customers as possible, thus capitalizing on the economy of scale associated with the networking equipment. The cost of resources that constitute a network scale up nicely, e.g., the cost per bit per second of transmission is vastly lower at higher bit rates. At lower bit rates, the decline of cost over time has not been spectacular over the past couple of decades.

The key raw material that powers the telecommunication networking gear is processors. The processor technology is driven by another empirical observation attributed to Gordon Moore and known as the Moore's Law. The Moore's Law states that the number of transistors on a chip will double every 18 months. The implication of this law is that the processing power will double every one and a half years at virtually the same cost.

The economics associated with the information network industry has got to be affected by the characteristics cited in the last two paragraphs, which are peculiar to this industry. Since pricing reflects the cost over a defined time horizon, it needs to reflect likely reduction in the cost of equipment along with the increased value to the customer and, possibly, the higher level of functionality demanded by the customer in the future. The combination of all these factors makes actual pricing of communication services as much an art as a hard science.

The network might be subject to geographical or political boundaries thus limiting the option of the network provider to capture all potential customers at the global level. Antitrust considerations might be applicable thus preventing telecommunication service providers from becoming an actual or de facto monopoly. However, the importance of communication has been globally understood by all the peoples of the world for well over a century. Communication among the people living in different countries is enabled through gateways or points of traffic exchange among cooperating service providers. National networks have evolved an efficient mechanism to interconnect and share resources and revenues to their mutual benefit and that of their customers. As an example, in the case of the

Internet, international communication is seamless and requires no additional effort by the communicating parties.

9.3 Economic Characteristics of an Information Network

Information networks are very different from other commodity producing or conveying facilities. We have mentioned earlier that there is an inherent economy of scale associated with a network. A single strand of optical fiber can convey up to hundreds of gigabits of information per second. Once a network is set up, the cost of operating it at or below its full capacity is more or less constant. This is in sharp contrast to an electric power generating facility (which also has a large capital cost associated with it) where the major component of the operating cost—the consumption of fuel—can be controlled to match the current demand. Because of these two factors—a high initial cost and the nearly constant cost of running the network (i.e., being independent of the traffic demand on the network)—the incremental cost of transporting information is close to zero. This condition holds as long as the network is operating below its rated capacity. When the incident traffic on the network reaches the network's specified capacity, the behavior of the network changes sharply. This behavior is manifested at the customer level by a sharp reduction in its performance and degradation in its throughput. Pricing is one mechanism to regulate the traffic offered to the network. The other mechanism would be to deny access to the network based on the state of the network.

In circuit switched networks such as the legacy telecommunication network, excess demand beyond the carrying capacity of the network is handled by making the dial tone unavailable to a caller or returning a fast busy signal if an intermediate trunk were busy. This has provided an effective way to handle congestion in the network. However, such a scheme cannot be easily implemented in a packet switched network such as the Internet. The Internet carries a fundamental premise that it would be always on. Excess demand in packet switched networks is handled by throttling the speed of communication that results in additional delay between the communicating parties. The control is thus exercised gradually. However, as shown in the earlier chapters of the book, increasing delays will eventually make the network unavailable for service.

Efficient markets are composed of suppliers and customers groups maximizing their utilities that arise out of production and consumption, respectively. Just like any other enterprise a telecommunications firm would like to maximize its profits. Profit is nothing but the difference between revenue and costs. A competitive market structure requires that firms determine their level of production at a point where marginal revenues equal marginal costs. Marginal revenue equals the revenue associated with one unit of output. Similarly, marginal cost is the cost associated with one unit of that same output. Marginal cost is also known as the incremental cost.

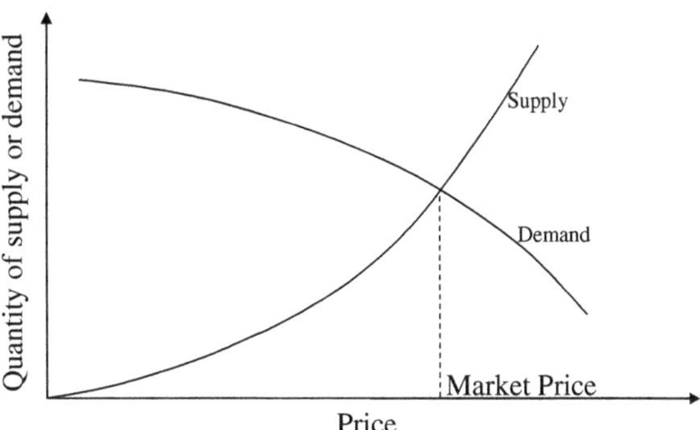

Fig. 9.1 Supply and demand curve

A similar situation exists from the consumer's perspective. Buying of a service or a commodity increases a customer's utility. In general, the utility curve is concave, meaning that the marginal increase in utility relative to the incremental expenditure incurred (or the marginal price paid) reduces as the amount of the service or the commodity purchased increases. The consumer will increase the amount purchased until the marginal utility equals the marginal price. At this volume of purchase, the interest of the consumer is optimized. The supply and demand curves as a function of the amount of goods produced by the supplier or consumed by the buyer are shown in Fig. 9.1.

As shown in Fig. 9.1, prices are defined by the intersection of supply and demand curves. From a consumer's standpoint, it occurs when the marginal utility equals the incremental price paid. The shape of the supply and demand curves vary from industry to industry. These curves for telecommunication services are markedly different from those applicable to typical commodities, for example, food or transportation. Since the cost associated with a telecommunications network is more or less constant, the marginal cost associated with a telecommunications network is nearly zero. This occurs until the network reaches a level of saturation beyond which point the performance degrades sharply and the service is severely affected. This characteristic is germane to all information networks and should become an important factor in the choice of technology, pricing, and regulatory framework associated with an information network.

The low marginal cost obviously breaks down when the telecommunication network reaches saturation. The service provided by a telecommunication network is throughput (or bits received by the addressee in a unit of time) against the bits required to be sent and entrusted to the network. Telecommunication users pay for service based on the throughput they receive assuming the quality of service is met. The bulk of the book has concentrated on evaluating the degradation in throughput as the number of nodes (or links) between the source and the destination increases. Since price is based on throughput, the findings reported in this book have emphasized the need for keeping the number of links between a source and the destination as few as possible.

The telecommunications network marketplace has three major stakeholders. We have already seen the impact of suppliers and consumers; each group is trying to maximize its profit or utility, respectively. The third important entity is the group of municipal, state or federal regulators. All three groups influence the actual price the customer pays for service.

One important reason to regulate telecommunications services is to ensure that there is adequate competition in the marketplace. Competition has been shown to result in innovation and, generally, lowering of price. Regulation has other functions as well. For example, regulators might require the very basic or essential services to be made available at a low cost. The regulator also might take into account interests of national security or policies of the government in determining functionalities associated with offered services and the associated tariffs to which all parties must adhere.

The regulator also has an important function from the perspective of economic theory. Acting on behalf of its constituency, the regulator's motivation is to maximize the social welfare. Social welfare is defined as the sum of the profits of all the suppliers and the surpluses of all the consumers. Surplus of a consumer is defined as the difference between the consumer's utility as a result of the service purchased and the price paid for the service. Under a perfectly competitive market scenario, it's possible that the tariff published by the regulator maximizes the welfare of the jurisdiction for which the regulator is responsible.

In the earlier part of the book, we have shown that packet switched networks are very sensitive to the number of hops as far as the throughput where a predefined quality of service is concerned. This finding nicely corroborates a corresponding recent finding applicable to circuit switched networks [122]. Reference [122] has produced a mathematical construct under which the call completion rate in a multi-hop circuit switched telecommunication network goes to zero under an inordinate traffic demand. In contrast, the results have also shown that for a single-hop network, the call completion rate never goes to zero, but saturates at a level consistent with the capacity of the link connecting the source and the destination. The desirability of having a low number of links between a source and the destination thus holds for both circuit and packet switched networks.

This book has addressed the performance and throughput associated with a network whose underlying switching fabric is based on packet switching. What we have shown in the earlier chapters of the book allows the service provider to price its services based on the cost associated with the number of links or, alternatively, provision the network in a manner such that the use of the number of links in serving traffic is minimized, thus creating a highly efficient network.

9.4 Major Components of a Telecommunications Network

Before we address the question of pricing end-to-end telecommunication services, it is important to model the network as a whole and understand the implications of the cost associated with their major components. Figure 9.2 is a diagrammatic

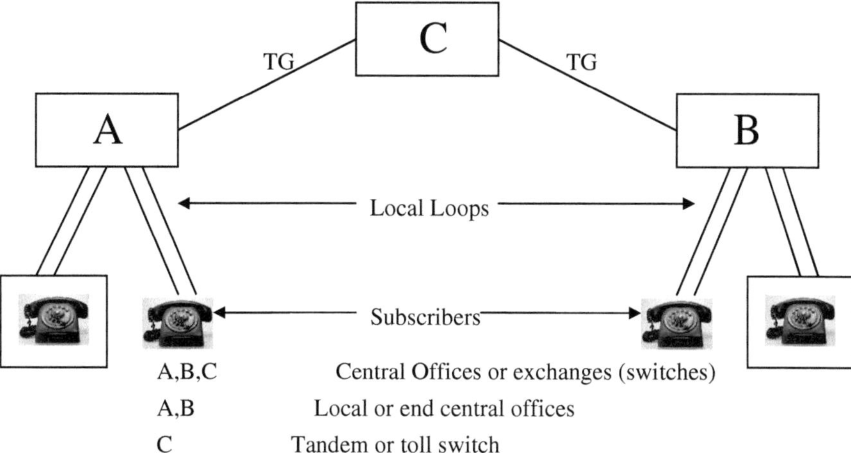

Fig. 9.2 The legacy telecommunication networking model

representation of the legacy telecommunications system. A major component of any telecommunication system is the local loop that connects individual subscribers to the telephone network. The local loop begins at the subscriber's premises and terminates at the subscriber's home central or switching office. This distance varies from a couple of miles to about ten miles or so. It has not been cost effective to multiplex signals from a number of subscribers to the first switching office; as such, each local loop caters to just one subscriber. Also shown in Fig. 9.2 are the switching offices, which might be the local switching office or a tandem switching office connecting two local or end switching offices, or another tandem switch. The switching office itself contains the switches that direct incoming traffic onto different trunk groups toward their destinations. The trunk groups are generally much longer in length and, unlike the local loops, are not permanently assigned to a specific user. The trunk lines that together form a trunk group have undergone major cost and performance improvements in the last couple of decades with the large-scale implementation of optical transmission technologies. Since trunk groups are not permanently assigned to a specific user, increasing their capacity has proven to be very cost effective. This is where the transmission technology has played its role in reducing the cost of long distance transmission by a significant amount over the last couple of decades.

A very rough assignment of relative costs incurred by the telecommunication service providers is as follows: local loop—40%; switching equipment—30%; and trunk or long distance transmission costs—30%. The local loop obviously presents a very large fraction of the overall costs and presents a formidable barrier to entry for a new service provider if it were required to build its own facilities. The Telecommunications Act of 1996 and its subsequent modifications require an incumbent service provider to open up these loops to its competitors for offering competitive services. This has enabled Internet Service Providers to spring up and offer Internet services using the existing local loops. VoIP services, where offered, ride over the Internet service as a data service. Largely attributed to regulatory reasons, data services and telephony services are treated very differently when it comes to pricing the services. The difference results in legacy telephony subsidizing data services that are offered over the Internet. From a regulatory standpoint, VoIP is treated as a data service thus influencing its price relative to legacy voice services. The following section shows how the resources of the telecommunication network offering legacy voice services as well as Internet services are tied together.

9.5 The Integrated Local Loop

Figure 9.3 shows the digital subscriber loop (DSL) that allows telephony and Internet access services to be integrated over a single pair of copper wires connecting an individual subscriber to the nearest central office. The copper wire was laid several years, possibly decades, ago when the need for the loop to carry information at up to several megabits per second was not comprehended. The local loops that carry DSL signals are upgraded or conditioned in order to have this capability. Such conditioning does not come cheap, and might require an investment of several thousand dollars per loop on the part of the service provider.

The DSL customer is equipped with a router provided by the service provider with the means to connect to it both the legacy phone equipment as well as data equipment, such as a PC. At the central office, the DSL is terminated in a digital subscriber line access module (DSLAM). The DSLAM splits and directs digital voice into the legacy circuit switched network that might be using a conventional digital switch or the asynchronous time multiplexed (ATM) switch. The data

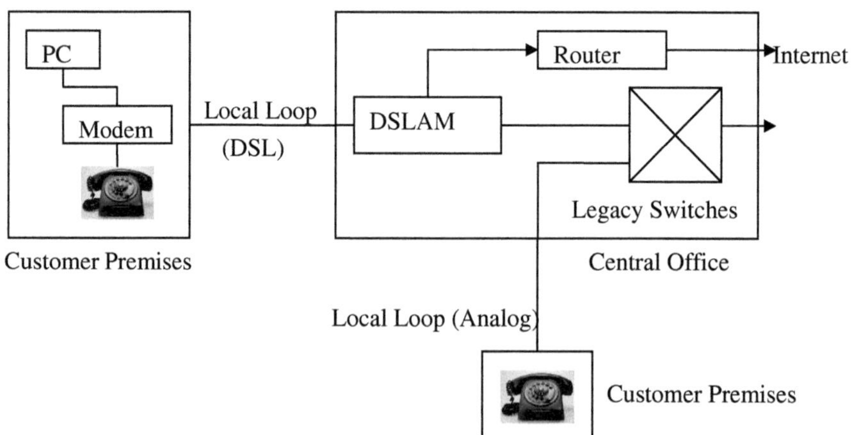

Fig. 9.3 Advanced digital subscriber loop

traffic is routed through a router to the Internet. Within the wide area network, voice and data travel separately to their respective destinations. However, in the access network they use the same medium. Internet Service Providers can lease just the data part of the access at tariffs largely determined by the telecommunications regulatory authorities. Given the current low rates at which the facility must be leased, there is simply no incentive for a competitive carrier to build its own access network.

9.6 Architectural Dissimilarities Between the Internet and the Legacy Network

The core of the Internet is a relatively simple packet transport device. Bits are packaged in packets that contain the addresses of the originator and the recipient of the information. The packets are steered by the routers that form the switching or routing points within the network. The Internet is oblivious as to the content of the packet itself. Higher-level protocols such as the TCP or UDP allow some discrimination as to the manner in which the packets should be treated on an end-to-end basis. The main feature of the Internet Protocol is that it allows the end points to exercise intelligence which is in sharp contrast to the legacy voice network where the end points are supposedly stupid and the intelligence is embedded into the core of the network. The architectural difference makes the legacy voice network extremely inflexible as far as the ability of a customer to develop applications for individual use is concerned. The opposite is the case with the Internet which easily allows development of individualized applications by its users.

The architectural dissimilarities between the two networks are driven almost entirely by the state-of-the art when the two networks evolved, respectively. Microprocessors were not invented when the architecture of the legacy voice network was nailed down. In addition, the price-performance characteristics of the electronic components and devices were entirely different when the architecture of the voice network was irrevocably defined. The deeply embedded intelligence in the legacy voice network was a result of the need to share intelligent devices among a number of customers, rather than making it available under the control of individual users, thus reducing its cost impact on the users. This made the voice network extremely inflexible when compared to the Internet from the customers' perspective.

Despite the architectural dissimilarities between the Internet and the legacy voice network, there is a fundamental unity among all forms and formats of information. Information has a common mathematical measure. This measure is applicable to all forms and formats of information and reduces any information to a common measurable denominator, called the bit. This unity among all forms of communication was not commonly understood until late in the evolution of electrical communication itself. The first electrical communication came about with the advent of the electric telegraph invented by Samuel Morse in 1844. It would take another 32 years before Alexander Graham Bell would file for the patent of the telephone. Separate networks were envisaged for the different formats of information in the early days of communication.

A unified mathematical theory of communication was formulated by Claude Shannon of Bell Laboratories in 1948. The theory led to the understanding of communication of information as one single function that could be addressed through one common infrastructure. Prior to the advent of the mathematical theory of communication, the fundamental unity among the different modalities of communication was not clearly envisaged. This resulted in some cases actually establishing separate networks for telegraph, voice, video, or data transmission. Increasing appreciation of the fact that all forms and formats of communication are fundamentally identical, combined with the rapid progress in digital signal processing, has led to what is now known as convergence. A variety of services can be offered over a converged network. A converged network would benefit from the economy of scale and the ability to create applications that integrate voice, data, or video functionalities. Different services, of course, require different levels of bandwidth. A reasonable approach would be to price the different services based on the network resources they actually consume.

This is not always the case, however. Regulatory intervention can distort the pricing picture by a considerable amount. One compelling example of this is the following fact. Legacy voice services offered over the legacy voice network today represents nearly 80% of the revenue of the telecommunication service providers. The remaining part of the revenue, about 20% of the overall revenue, is through the Internet services. Even so, some 80% of the network bits can be attributed to data communication or bits related to offering non-voice services. In other words, the legacy telephone revenue is subsidizing the Internet services today.

Cross-subsidization of services or product groups is not an unknown phenomenon in businesses. Usually, a legacy service that is known to disappear in the future does not require new investments while producing revenues based on past investments. Such services or products are known as cash cows in the parlance of economists and are used to subsidize new services or products in their initial phase of development.

9.7 Access Network as the Bottleneck

The access portion of the telecommunication network is potentially the largest element of its cost structure. In the most general case, the access network is a resource for use by a single subscriber. Therefore, it has not benefitted from the tremendous leaps that transmission technology has enjoyed over the past couple of decades. The reason is that newer developments have allowed the cost per bit to decrease sharply at higher bit rates, but the scaling on the lower side of the scale has not been spectacular at all.

The result of this discrepancy is the fact that service providers that own facilities do not have a clear motivation to invest in the access part of the network, especially when these accesses must be made available to their competitors at fictitious costs. The reason we call the current means of pricing the access cost as fictitious is that whenever a service provider owns transmission facilities in order to provide its customers an integrated service, it's not clear how one can isolate that portion of its revenue that corresponds to raw transmission alone [123]. This is in spite of the fact that the judicial systems and the Federal Communication Corporation have done as good a job as can be expected with their limited vision of technology and its future evolution.

By almost any account, the regulatory framework as practiced in the United States over the past couple of decades has not functioned well as far as the access network is concerned. The United States has had the privilege of being the number one country in the telecommunication services. This is no longer the case and past regulatory interventions in the pricing of the access network is, possibly, the number one cause.

In the final analysis, different technologies that perform the same function should not be priced differently. The artificial separation from a regulatory perspective between the bits that belong to the legacy voice communication and data that travel over the same access pipe is not sustainable over the long term in a free market system. Even so, the regulatory authorities in the United States and several other countries have chosen to practice it creating an opportunity of arbitrage in the marketplace. VoIP obviously benefits from this arbitrage.

9.8 Pricing of Internet Services

The price associated with a minute call from the legacy phone anywhere in the US varies between 5 and 10 cents per minute. We shall take the lower price of 5 cents per minute for comparison purposes. For regulatory as well as reasons attributed to customer preferences, Internet service offerings are not measured, and therefore can be used by the subscriber with a duty factor that can approach 100%. Let's take the figure of $20/month for Internet services at 1 Mb/s. Assuming that VoIP services over the Internet can offer voice service at the nominal rate of 20 kb/s, each Internet access connection is capable of offering 50 simultaneous voice connections. Since there are 43,000 min in a month, there could be a maximum of 43,000 × 50 call minutes that can be serviced by an Internet access line over the period of a month against the cost of $20. Using 5 cents as the price of the call, Internet-based VoIP has the potential of generating revenue of $21,500. The motivation of a VoIP service provider to establish telephone services over the Internet with no stake in transmission facilities infrastructure is thus a compelling business proposition.

There is little doubt that this massive divergence between price and cost presents a tremendous opportunity for arbitrage. VoIP services are thus strongly propelled by the subsidization that the Internet provides.

The cross subsidization opportunity that the Internet offers is not limited to the above discrepancy. As it happens, the Internet-based VoIP services are classified by the Federal Communications Commission in the United States as Information services as opposed to the legacy voice services. As information services, VoIP offered over the Internet escapes the universal tax that can be as much as 9%. Unquestionably there will be a strong motivation for VoIP to continue having a compelling growth at the cost of legacy voice services in the foreseeable future.

9.9 The Overall Impact on VoIP Pricing

Broadly speaking, the market price of telecommunications service offerings is based on three major factors: the market itself (which includes consumers, suppliers and competitors), impact of evolving technology, and regulatory hurdles. The market thus includes the behavior of customers and competitors. The anticipated evolution of technology itself must be considered into pricing, since pricing of a service is based on the totality of costs over a period of time, usually the anticipated lifetime of the cost components that are reflected in pricing the service. Regulatory factors generally shift the cost of one element of service to another or from one market segment to the other.

The telecommunications marketplace shows a high sensitivity to cost. The low cost of the VoIP services will continue to increase the interest of customers in VoIP. Technological factors further bode well for the VoIP services because the reduction in the cost of electronic and photonic components will more immediately and directly affect the cost of offering VoIP services instead of in the pricing of legacy voice services. One compelling example lies in the fact that advancements in codecs will directly lead to bandwidth reduction for VoIP services. On the other hand, since the legacy voice network is built around a fixed bandwidth of 64 kb per second, such advancements will have no effect on the cost factors that affect conventional voice offerings.

The impact of regulatory factors has been considered earlier in this chapter. The regulatory direction now clearly allows the subsidization of data services—and VoIP is treated as a data service—at the cost of legacy voice services.

The result of all the above factors clearly points toward VoIP gradually, if not spectacularly, replacing the conventional voice services over a period of a decade or more. In addition to the factors mentioned above, it is generally understood that a singular network based on packet switching will allow the development of services that could not be economically offered over interconnected networks using disparate technologies. It would be worthwhile speculating on the likely path that the increasing integration of legacy and evolving technologies will take over a period of time. This is considered in the next section.

9.10 The Likely Evolution Toward a Single Network

The Internet and the legacy telephone network are significantly different in their architecture as well as in their business philosophies. Perhaps the most significant architectural difference between the two networks arises from the fact that there is the physical separation of signaling and payload information in the legacy network. The motivation for VoIP arises, among other factors, by the desire to integrate the functionalities of voice, data, and video into one single integrated network. The integrated network will benefit from the economy of scale as well as offer limitless opportunities for creating applications that will combine the functionalities of the three modalities of information, namely, voice, data and video.

The simplest level of physical integration between the two networks can be achieved by substituting the trunks in the legacy network with IP-based transmission. The network interfaces at the two ends of the emulated trunk group are not affected, and no other changes in the legacy network need to be made. The network operating in this mode will function as a 64 kb/s circuit switched network and the signaling system remains intact without being affected in any manner. The only motivation to implement such architecture will be based on a possible reduction in the cost of trunking by replacing a TDM-based trunk group with IP-based transmission. This integration is shown in Fig. 9.4. In this figure, A and B are two end switching systems connected by a tandem switch C. Interfaces to the three switches remain unchanged even though the trunk groups are replaced by high-speed transmission links.

Figure 9.4 does not reflect IP telephony or Voice over IP. Implementation of VoIP will require that the entire legacy network between the end switching offices A and B is replaced by the Internet. This can be achieved by transmitting the voice payload over the IP network. Furthermore, the signaling associated with the legacy network, based on the signaling system 7 (SS7), must be converted to the IP format and travel over the IP network. Packet switching can offer significant savings in the transmission cost by replacing the digital or G.711-based (uncompressed) voice signals with the G.723 signals using codecs that compress and packetize

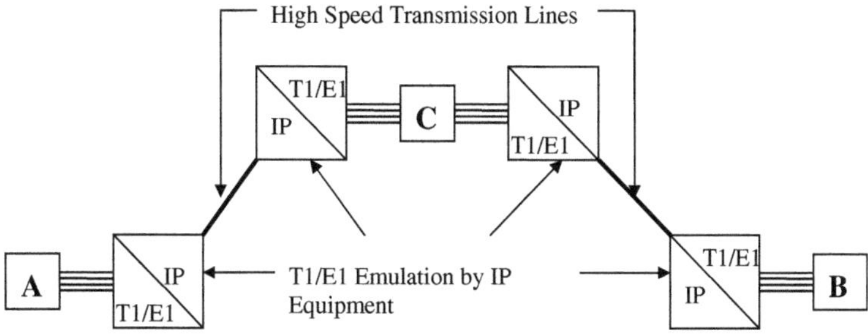

Fig. 9.4 An IP-based trunking scheme for the legacy network

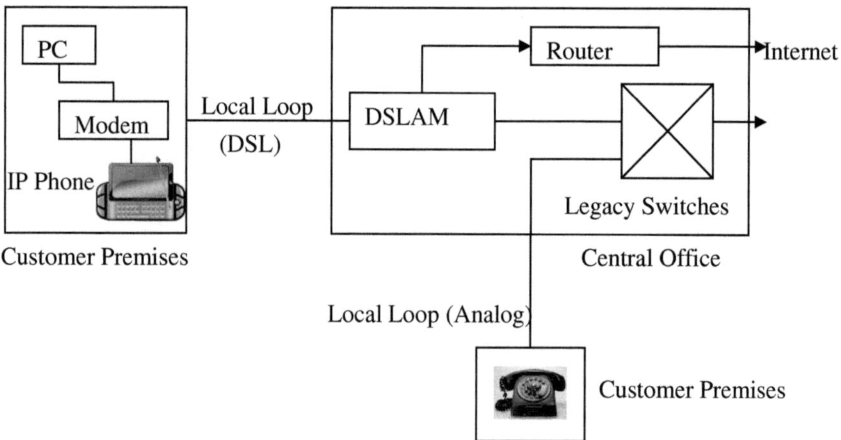

Fig. 9.5 VoIP using an integrated access scheme

voice signals. For reasons of real-time implications in interactive voice, the voice
packets would need to use the UDP instead of the TCP at Layer 4. Additionally,
the use of RTP or Real Time Protocol will have to be used as well to manage
effectively the integrity of interactive voice.

We offer two scenarios here. In the first scenario, the user retains the traditional
telephone instrument and the conversion of voice signals and signaling information into the IP packet format is performed at the central office. The user can thus
continue to dial using the traditional touch tone pad; however, this signaling
information will have to be extracted at the central office and converted into the
SIP signaling scheme in order to comply with the IP packet format. Over the
access network, data and voice travel in different frequency spectra and are, thus,
functionally separated even though they achieve physical integration through DSL.
This scheme is shown in Fig. 9.5. Beyond the router shown in the central office,
voice and data are integrated in the IP network and share its resources.

The legacy access scheme discussed above and shown in Fig. 9.5 can be
replaced by the IP access scheme that can have a higher level of integration in
the access system. Such a scheme can achieve the generation of the SIP signaling information right at the user's premises along with the generation of voice
packets using the IP format. Admittedly, the equipment owned by the user will
have to deal with dialing in the conventional E.164 scheme as well, using a
touch tone pad since not all destinations will have a SIP-like address. The
schematic diagram for this highest level of integration will look similar to that in
Fig. 9.5, except that the local loop will function as an IP-based transmission
medium where the voice, data or video transmission packets are not functionally
separated or assigned bandwidths not accessible to the other modalities of
communication.

9.11 Conclusions

Over the past several years, VoIP services have shown increasing acceptance coupled with reducing price. This chapter has captured some of the economic reasons for VoIP's growing popularity and ubiquity. We have first discussed some general characteristics of information and how information based services are different from other services or commodities. The VoIP service benefits from being classified as a data service as opposed to a voice service thus enjoying relief from regulatory hurdles that have been placed on voice services. Furthermore, VoIP uses Internet for its wide area transport. The Internet has a very low transmission cost per bit because of the economy of scale and use of the packet switching technology. Finally, as the telecommunication network evolves toward a single structure, VoIP will be able to offer superior characteristics as far as integration with other data and video services are concerned.

Chapter 10
A Network Based Authentication Scheme for VoIP

Abstract This chapter presents a VoIP networking solution that incorporates network-based authentication as an inherent feature. The proposed authentication feature introduces a range of flexibilities not available in the PSTN. Since most calls will likely terminate on the network of another service provider, a mechanism using only networks which can mutually authenticate each other to insure authentication across networks is also presented [124].

10.1 Introduction

Telephone systems have evolved over the last century. This evolution has moved from analog to digital systems and from using the circuit switching technology to packet switching systems. More recently, the potential benefits of converged networks in addition to lower costs for the customer has led to increasing use of Voice over IP [104]. This chapter focuses on just one aspect of Voice over IP, namely, mutual authentication of the calling and called parties. Mutual authentication between the caller–callee pair has been a distinguishing characteristic of the PSTN for over a century, but more so over the last 20 years or so when several applications use authentication provided by the telephone network for identification of the parties for sensitive transactions.

Voice over IP applications are generally designed to function over the global Internet, although such solutions can be offered over private IP networks, such as enterprise networks. Instances of violations of security over the Internet are common occurrences that affect individuals and businesses as well as government operations. Voice over IP has not yet suffered many security violations, but the potential for attacks on security is truly large. Possibly the mass of end points

P. K. Verma and L. Wang, *Voice over IP Networks*, 111
Lecture Notes in Electrical Engineering, 71, DOI: 10.1007/978-3-642-14330-4_10,
© Springer-Verlag Berlin Heidelberg 2011

connected to VoIP today is below the threshold that would attract miscreants. However, faking the identity of the caller can be easily accomplished using the Internet. This can result in the unsuspecting callee passing sensitive information to the calling party. Additionally, the negative impact of SPAM over Internet Telephony can be easily comprehended.

What level of security would two persons conversing over the emerging converged network need? There are three primary properties of secure communication: secrecy, authentication, and integrity. For real-time voice communication among known individuals, authentication is guaranteed by their mutual recognition of each other through what is broadly known as speaker recognition. However, there are two caveats to relying on this as an exclusive authentication mechanism. First and foremost, authentication needs to take place prior to the commencement of communication, because it indeed might form the basis of call acceptance. Second, the caller might be led to a machine such as a human–machine interactive system where speaker recognition is crucial to the task being performed.

Authentication, as a generally available feature, is more important than secrecy as far as voice communication is concerned. While authentication can be relegated to the end points (just as secrecy or integrity might be), it is considered extremely important from the security of the network's perspective to identify and log the party that initiates the call. Most of the security issues in the Internet exist because of its inability to identify positively the party that initiates the call. This enables an intruder to escape detection by the network.

With the increasing demand for the VoIP application, Session Initiation Protocol (SIP) [25] has been developed by IETF for establishing real-time calls and conferences over IP networks. SIP is an end-to-end client–server signaling protocol to establish presence, locate users, set up, modify and tear down sessions. As a traditional text-based Internet protocol, it resembles the hypertext transfer protocol (HTTP) and simple mail transfer protocol (SMTP). Not being limited to IP telephony, SIP messages can convey arbitrary signaling payload, session description, instant messages, JPEGs, and any MIME type. SIP uses session description protocol (SDP) [51] for media description.

Despite many advantages of SIP, it is subject to various kinds of Denial of Service attacks. User authentication is becoming more and more important to prevent unauthorized users from using someone else's identity to fool other users or accounting and charging systems. Furthermore, mutual authentication of the calling–callee pair is an essential requirement for the successful execution of several applications executed over a voice platform.

This chapter presents an authentication scheme that can be implemented within the SIP security framework. It is organized as follows. In Sect. 10.2, the importance of authentication is described. Current methods for authentication are discussed in Sect. 10.3. Section 10.4 outlines new requirements of both users and networks. Our proposed scheme meeting these requirements is addressed in Sect. 10.5. The chapter is concluded in Sect. 10.6.

10.2 Authentication

10.2.1 Need for Authentication

This section presents an expanded notion of authentication for both the calling and the called parties. It is anticipated that positive identification of the calling party will become an essential requirement for the callee to engage willingly in a conversation or engage in an application to be executed over a voice platform. At the present time, the calling line identification (CLID) serves this purpose. In VoIP, the CLID can be easily manipulated and, therefore, will be increasingly suspect. Likewise, the calling party needs to be sure of the identity of the callee as well, e.g., when a bank employee or a credit card company calls an account holder to inquire about a specific transaction. This would imply that the instrument terminating the call and/or the party receiving the call is positively identified by the network. For maximum protection, both the calling and the called parties might need to be additionally identified using a biometric contrivance.

10.2.2 Mutual Authentication Versus Network Authentication

Authentication can, in general, be performed either on an end-to-end or identification by the network at points of ingress/egress basis. A case for the latter is made in this chapter. If the network were to authenticate an end user, the total number of authentication functions needed would be a maximum of n versus $n(n - 1)/2$ for the other case. End-points engaged in multi-party communication would need to identify themselves just once as opposed to each end-point identifying every other end-point on a pair-wise basis. Higher speed of operation as well as a large reduction in computational overhead and key management issues can be anticipated as a result of network based authentication.

A prerequisite for the network-based authentication is, of course, the existence of the network as a trusted agent of each end-point. In general, such a notion is opposite to the concept of the Internet where the network can be easily fooled as far as the identity of an endpoint is concerned. In fact, this lack of ability has resulted in the growing threat of hackers of the Internet. There is a class of users for whom anonymity must be guaranteed, assuming the party they wish to communicate with honors such anonymity. In order to assure anonymity, in our proposed scheme the network does not pass the identity onto any other user even though it has verified the credentials of the specific user wishing to communicate. Exceptions must also be granted for a caller in emergency, possibly through the intervention of a live agent. The scheme proposed in this chapter accommodates the anonymity needs of the caller and callee independently while recognizing that communication is not guaranteed among incompatible end points.

Table 10.1 Comparison of four authentication schemes

Scheme	Key	Type	Application	Flaw
Digest	Pre-shared key (PSK)	One-way challenge-response	END-to-END	Integrity and confidentiality not support
IPSec	PKI and PSK	Network layer	HOP-by-HOP	Considerable overhead
TLS	PKI	Transport layer	HOP-by-HOP	Over TCP not UDP
S/MIME	PKI	S/MIME contents	END-to-END	Large size message

10.3 Comparative Analysis of Existing Authentication Schemes

There are four commonly used solutions for authentication in the SIP framework: digest-authentication, IPSec, TLS and S/MIME, which are analyzed and compared in Table 10.1.

Digest authentication [125] is a challenge–response based scheme to offer one-way authentication to the first requested server, while the intermediary servers have no idea whether the user is verified or not. It gives opportunity to a malicious user to masquerade in the "next-hop," and make it vulnerable to spam, man-in-the-middle and DoS attacks. In addition, adopting the digest authentication scheme in an IP telephony system only authenticates the communication device lacking relationship between its user and the SIP URL.

IPSec provides security services at the IP layer by enabling a system to select required security protocols, determine the algorithms to use for services, and put in place any cryptographic keys required to provide the requested services. IP Encapsulating Security Payload (ESP) [126] in the tunnel mode is preferred for IP tunneling across the Internet, although it involves substantial overhead. With IPSec used in tunnel mode, payload efficiency (ratio of payload to total packet size) of a 40-byte VoIP packet drops from 50% to <30% [127], since the RTP/UDP/IP header is 40 bytes for IPv4.

Transport layer security (TLS) [128] provides a reliable end-to-end secure channel over connection oriented protocol. Both ends of the channel are identified by X.509 certificate [129] exchange. Making use of TLS to secure SIP signaling is transparent, which allows a signaling message at the application layer to be encrypted by TLS and then transferred through the TCP connection. If a TLS connection is requested, a SIP Secure URI (SIPS) is used. TLS is impractical to deploy in a wide area network since the TLS is built upon connection-oriented TCP protocol, restricting itself to limited applications, while most VoIP applications offer a continuous stream of RTP/UDP/IP packets. Furthermore, if one hop along the path does not support TLS, the transit trust loses its meaning [130].

Secure multipurpose internet mail extension (S/MIME) [131] has been developed for electronic messaging applications to provide origin privacy and message integrity. Conventionally, S/MIME used in SIP is for user-to-user or user-to-proxy authentication. In our proposed scheme, S/MIME will be adopted to encrypt the entire SIP message for hop-by-hop authentication instead of using an external protocol such as IPSec or TLS. Our proposed scheme could not be replaced by digest authentication, which needs a pre-shared secret key between all users and between all users and proxy servers.

Some other solutions are proposed in the literature based on combinations of these four schemes. Johan Bilien has recommended solutions to secure VoIP using S/MIME, TLS and MIKEY for SIP signaling to provide end-to-end authentication and session key distribution, using SRTP on payload to protect voice media [132, 133]. NTT Network Service Systems Laboratories has proposed two approaches to provide security on both SIP signaling and media streams for end-to-end communication in different scenarios [134]. In the following section, the enhanced view of authentication in VoIP is proposed.

10.4 Proposed Requirements of Authentication

Our proposed scheme offers the caller and callee options for authentication independently. In other words,

- the caller might choose to remain anonymous to the callee (not network).
- the caller might choose to reject anonymous calls.
- both caller and callee can choose protection against eavesdropping, i.e., invoke encryption of the message.
- callee can ask for more details related to caller's identity leading to a higher level authentication, such as website, birthday, or last transaction.

It is also important that the proposed network based authentication scheme accommodate authentication across one or more networks.

10.5 Proposed Schemes for Authentication

10.5.1 Proposed Scheme

Our proposed scheme assumes intra-networks with trust relationships between hops. S/MIME is applied to the SIP message including headers and bodies, which requires each proxy server to decrypt and encrypt the SIP message using respective public/private key pairs of the two communicating parties. It means that both user-to-proxy and proxy-to-proxy authentications are based on S/MIME to continue passing the transit trust one by one. Figure 10.1 shows the overall operation and flow of messages within and between the two parties, which are the User Agent and ingress proxy server. Details are as follows:

1. The signature of the SIP message is created by encrypting the digest of the message with the caller's private key. The proxy server authenticates the caller using the caller's public key.
2. The caller randomly generates the session key protecting voice data against eavesdropping if so desired. Options are available to the caller/callee as described

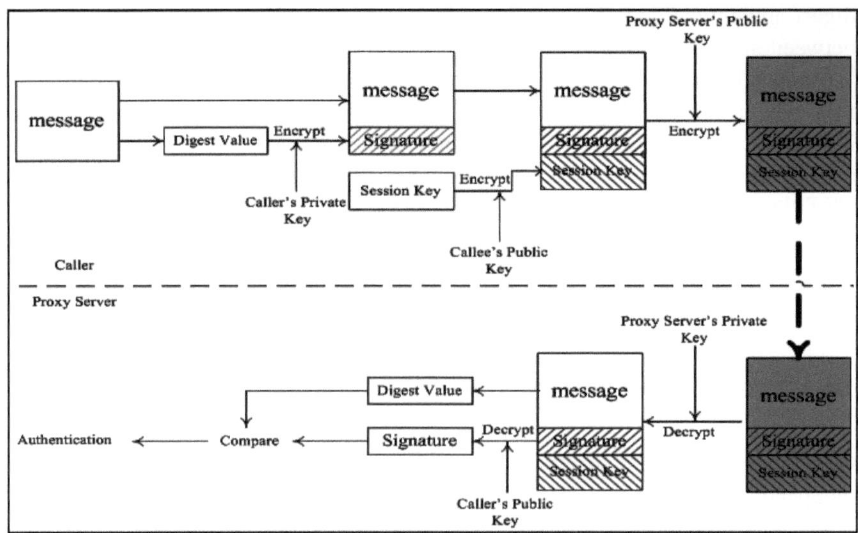

Fig. 10.1 S/MIME INVITE message

in Sect. 10.5.2. The session key is generated by applying 3DES on a pseudo-random number and securely sent to the callee by encrypting it with callee's public key, which can be retrieved from X.509 Certificate hierarchy [135].

3. The SIP message, caller's signature and the encrypted session key are concatenated and encrypted using the public key of the proxy server.

4. The proxy server decrypts the received message with its private key and the signature with caller's public key. It compares the recomputed digest value of the message with the decrypted signature. If they are matched, it is a valid signature and the caller's identity is verified. At the same time, authentication of the proxy server is also proved since only it has the private key to get the message. Note that the session key will not be disclosed to the proxy server because only the callee knows his/her private key.

Each proxy server along the signaling path executes the same authentication procedure as depicted above, but with its own private key for signature and the public key of the next-hop server for encryption, resulting in end-to-end authentication. The callee finally decrypts the session-key with his/her private key. The session key will be applied for protection of media flow; however, it is not a requirement for end-to-end communication. The two users' preferences for anonymity or desires for confidentiality are indicated in the matrix discussed below.

10.5.2 Enhanced Performance

In order to meet the users' needs as indicated in Sect. 10.4, one new header is introduced in the SIP signaling message call-type to offer more options.

Table 10.2 Caller/Callee option

Anonymity (First digit)		Confidentiality (Middle two digits)		
Caller	0 No	00	Either	Reserved (last digit)
	1 Yes	01	Yes	
Callee	0 Accept all	10	No	
	1 Reject anonymous calls	11	Caller Reserved Callee More information required for authentication	

Table 10.3 Call-type

Anonymity	Rejected call		More information required	Confidentiality
	Anonymous call	Confidentiality not matched		
1XXXXXXX	1XXX1XXX	X01XX10X	XXXXX11X	X00XX00X
				X00XX01X
		X10XX01X		X01XX00X
				X01XX01X

The call-type header is defined as an 8-bit field. The first/last four bits are set respectively by caller/callee to indicate their preference for making/receiving an anonymous or confidential call. In addition, the callee has one more option to ask for further verification of the caller's identity with more information. Each four-digit group is assigned as below in Table 10.2.

Although the proposed system requires only 3 bits, a 4-bit field is recommended for the caller/callee with the last bit reserved for future use. As Table 10.2 shows, the caller has six options while the callee has eight options. There are 48 valid combinations for the network to decide the properties of the call to be established.

For anonymity, the proxy server that receives the first INVITE message directly from the caller, checks the first bit. If it is 1, the server will remove from and contact headers from the decrypted message and add appropriate information in the via header field; next, the server signs and encrypts the modified message, and then passes it onto the next hop.

For more information required, the proxy server that receives response directly from the callee, checks all the 8 bits in the way depicted in Table 10.3. If the 8-bit matches XXXXX11X, this proxy server returns 401 unauthorized including the call-type and indicates that the request requires additional authentication. Once 401 unauthorized is received by the caller, he will generate a new INVITE message with more information and send it to the callee again in a way similar to digest-authentication. After the additional authentication, the actual call will take place according to the confidentiality preferences of the parties, if they match as indicated in Table 10.3.

10.5.3 Inter-network Authentication

Making VoIP calls crossing different networks needs network-to-network authentication, which is performed at the gateway between networks. It is similar to hop-by-hop authentication described above. The gateway will attach the sending network's signature, and then encrypt the message together with the signature and wrapped session key using the communicating network's public key obtained from X.509 Certificate. The only difference is that this authentication procedure moves to the gateway, which keeps the public/private key pair of the network.

10.6 Conclusion

This chapter has summarized and compared various solutions to SIP signaling authentication. After reviewing current proposals, a scheme for authentication is proposed to be executed by each hop: user-to-server, server-to-server or network-to-network. By chaining trust among SIP components across the trusted network, end-to-end authentication is realized. In addition, advanced level of authentication service is offered for users in our scheme through options of anonymity, confidentiality and additional authentication, if required.

Chapter 11
Conclusions and Future Work

This book has studied the performance of the VoIP networks from three aspects, Quality of Service, price, and security. As a real-time application, VoIP networks are challenged by the legacy voice quality as exhibited by the traditional PSTN. This book has introduced the notion of upper bound of delay and jitter, termed threshold, to characterize the QoS of VoIP traffic. Traffic that suffers a delay/jitter higher than the bound is considered lost and does not constitute effective throughput. This book has presented the impact of the threshold delay/jitter on the resource consumption under different queuing models which have proven to represent adequately traffic for VoIP. The learning from the analysis has been extended and applied to assessing the impact of risks constituted by a number of transportation channels where the risk associated with each channel can be quantified by a known distribution. Finally, this book has also presented an authentication mechanism using which networks can mutually authenticate each other to afford the possibility of authentication across networks.

In the first part of this book, the average delay and the bounded delay have been compared in characterizing both the QoS and network management aspects. Two different queuing models have been evaluated, M/M/1 and M/D/1. For the M/M/1 model, a single-hop network can achieve a 100% throughput, if the upper delay bound goes to infinity, while the maximum throughput of the n-hop network can only reach $\rho^n\%$. For the M/D/1 model, a similar result is found in that the throughput performance of the multi-hop network continues to deteriorate as the number of hops increases; the needed capacity of both the single-hop and the multi-hop networks can be analytically evaluated from the closed form solution presented. The results obtained can be used in scaling resources in a VoIP network for different thresholds of acceptable delays. The notion of the upper bound delay has been also extended to jitter which is, potentially, the largest source of degradation in the quality of voice in VoIP systems. This book has also provided a way to compute the traffic handling capability of a multi-hop resource constrained network

P. K. Verma and L. Wang, *Voice over IP Networks*,
Lecture Notes in Electrical Engineering, 71, DOI: 10.1007/978-3-642-14330-4_11,

under a defined limit of end-to-end jitter. The presented analytical solutions help scale up the resources in the network as the number of hops increases.

The second part of this book has proposed a new pricing scheme based on the cost of lost opportunity vs. the cost of consumption of resources. The notion of a unit price has been developed. Separate results are derived for single-, two- as well as multi-hop networks. It has been observed that the unit price for two-hop traffic is more than four times the unit price for single-hop traffic if the performance were to remain the same. The examples given in this book show that both the ratios of the capacity per each link and the unit price of a two-hop network, relative to a single-hop network, decrease as the threshold delay increases. A similar pricing solution for a multi-hop network is also presented which depends on the number of hops as well as on the relative change in capacity of each hop compared to the single-hop case with identical QoS. This book has appropriately priced VoIP services at equitable levels that are consistent with the resources consumed in order to achieve the contracted QoS.

Extending the techniques developed in the previous chapters, this book has evaluated the probability that the cumulative risk of a number of transportation channels in series is bounded to a specified value. In particular, the risk associated with the flow of containerized traffic in a cascade of channels with diverse risk characteristics has been used as an example. The book has developed a closed form solution in terms of the characteristics of the constituent channels with dissimilar risk characteristics. This approach can be used to shape the risk characteristics of individual channels through additional investment in order to maximize the impact of such investments.

The third part of this book has considered network security solutions in VoIP applications. The potential for attacks on security is large in the VoIP environment. For the network authentication, this book has presented a VoIP networking solution that incorporates network-based authentication as an inherent feature and introduces a range of flexibilities not available in the PSTN. After reviewing current proposals of SIP signaling authentication, a new authentication scheme has been developed based on the chaining of trust among SIP components across the trusted networks, realizing end-to-end authentication. In addition, an advanced level of authentication service is offered through options of anonymity, confidentiality and additional authentication, if required.

Future work would potentially include a further generalization of the solutions provided in this book from different standpoints. Additional queuing models such as M/G/1 and G/G/1 can be considered for analysis to improve the optimum channel capacity allocation in a more generalized manner. It is also possible to define the packet size ranges based on actually measured statistics of traffic that traverses the Internet. Furthermore, the independence assumption used in this book can be weakened to include self-similar traffic. For implementing VoIP security, authentication can be done via biometric features such as finger print, eye scan, voice spectrogram, etc. The authentication mechanism presented in this book can be further augmented to address other security issues such as confidentiality and non-repudiation.

References

1. Goode, B.: Voice Over Internet Protocol (VOIP). *Proc. IEEE* 90, 1495-1517 (Sept. 2002)
2. Bonenfant, P.A., Leopold, S.M.: Trends in the U.S. communications equipment market: a Wall Street perspective. in 44, 102–108 (Feb. 2006)
3. Morgan Keegan & Co., Proprietary Equity Research, (2005)
4. Cherry, S.: Seven myths about voice over IP. in *IEEE Spectrum* 42, 52–57 (March 2005)
5. Wallace, K.: Authorized Self-Study Guide Cisco Voice over IP (Cvoice). Cisco Press, Indianapolis, IN (2006)
6. Thomsen, G., Jani, Y.: Internet telephony: Going like crazy. in *IEEE Spectrum* 37(5), 52–58 (May 2000)
7. Stallings, W.: Data and Computer Communications. Prentice-Hall, New Jersey, USA (1997)
8. Maresca, M., Zingirian, N., Baglietto, P.: Internet protocol support for telephony. ProcProc. *IEEE* 92, 1463-1477 (Sept. 2004)
9. Chong, H.M., Matthews, H.S.: Comparative analysis of traditional telephone and voice-over-Internet protocol (VoIP) systems. *Electronics and the Environment, IEEE International Symposium*, May pp. 106–111 (May 2004)
10. Hamdi, M., Verscheure, O., Hubaux, J.-P., Dalgic, I., Wang, P.: Voice service interworking for PSTN and IP networks. IEEE Commun. Mag. **37**(5), 104–111 (1999)
11. Hardy, W.C.: QoS Measurement and Evaluation of Telecommunications Quality of service. John Wiley & Sons, West Sussex, England (2001)
12. Jha, S., Hassan, M.: Engineering Internet QoS. Artech House, London, UK (2002)
13. Armitage, G.: Quality of Service in IP Networks. Macmillan Technical Publishing, Indiana, USA (2000)
14. CT Labs, Inc.: Speech Quality Issues & Measurement Techniques Overview, Revision: 10-23-2000 CJB, 2000. http://www.ct-labs.com/Documents/Speech_Quality_Testing.pdf
15. Morrow, M., Sharma, V., Nadeau, T.D., Andersson, L.: Challenges in Enabling Interprovider Service Quality in the Internet. in *IEEE Communications Magazine* 43, 110–111 (2005)
16. Christos Bouras and Afrodite Sevasti: Pricing QoS over transport networks Internet Research, 14(2), 167-174 (2004)
17. Odlyzko, A.: Internet pricing and the history of communications. *Computer Networks* 36(5), 493-517
18. Qu, Y., Verma, P.K.: Notion of cost and quality in telecommunication networks: an abstract approach. IEEE Proc-Commun **152**, 167–171 (2005)
19. Walsh, T.J., Kuhn, D.R.: Challenges in securing voice over IP, in *IEEE Security & Privacy Magazine* 3, 44–49 (May-June 2005)
20. Sun, L.: Speech Quality Prediction for Voice over Internet Protocol Networks, Ph.D. Thesis, University of Plymouth, January (2004)

21. Schulzrinne, H.: http://www.cs.columbia.edu/~hgs/internet/
22. Schulzrinne, H., Casner, S., Frederick, R., Jacobson, V.: RTP: A transport protocol for real-time applications, IETF RFC 1889, 1996. ftp://ftp.ietf.org/rfc/rfc1889.txt
23. Kurose, J.F., Ross, K.W.: Computer Networking: a Top-down Approach Featuring the Internet. Prentice Hall, New Jersey, USA (2006)
24. ITU-T Recommendation H.323: Packet-based multimedia communications systems, International Telecommunication Union, (1997)
25. Rosenberg, J., Schulzrinne, H.: Camarillo, Johnston, Peterson, Sparks, Handley and Schooler, SIP: Session initiation protocol v.2.0, IETF RFC 3261, (2002)
26. Arango, M., Dugan, A., Elliott, I., Huitema, C., Pickett, S.: Media gateway control protocol (MGCP) Version 1.0. IETF RFC 2705, (1999)
27. Greene, N., Ramalho, M., Rosen, B.: Media gateway control protocol architecture and requirements. IETF RFC 2805, (2000)
28. Cuervo, F., Greene, N., Rayhan, A., Huitema, C., Rosen, B., Segers, J.: Megaco Protocol Version 1.0. IETF RFC 3015, (2000)
29. Wright, D.J.: Voice over Packet Networks. John Wiley & Sons, Chichester, West Sussex, England (2001)
30. Schulzrinne, H.: IP Networks. http://citeseer.nj.nec.com/schulzrinne00ip.html
31. ITU-T Recommendation G.114 'One-way transmission time' (1996)
32. Stability and Echo, CCITT Recommendation G.131, (1988)
33. Hassan, M.: Internet Telephony: Services, Technical Challenges, and Products, *IEEE Communication Magazine* (Apr. 2000)
34. Bai, Y., Ito, M.R.: QoS Control for Video and Audio Communication in Conventional and Active Networks: Approaches and Comparison. *IEEE Communications Surveys & Tutorials* 6(1), First Quarter (2004)
35. De Vleeschauwer, D., Büchli, M.J.C., Van Moffaert, A.: End-to-End Queuing Delay assessment in Multi-service IP Networks. Journal of Statistical Computation and Simulation 72(10), 803–824 (2002)
36. Sreenan, C.J., Chen, J., Agrawal, P., Narendran, B.: Delay Reduction Techniques for Playout Buffering. IEEE Transactions on Multimedia 2(2), 88–100 (2000)
37. Wang, R., Hu, X.: VoIP Development in China. Computer 37(9), 30–37 (2004)
38. Bai, Y., Ito, M.R.: A Study for providing better quality of service to VoIP users. *IEEE Proceedings of the 20th International Conference on Advanced Information Networkinng and Applications (AINA'06)* vol. 1, pp 799-804
39. Boutremans, C., Le Boudec, J.Y.: Adaptive joint playout buffer and FEC adjustment for internet telephony. *IEEE INFOCOM'2003*, San-Francisco, USA, (Apr. 2003)
40. Jiang, W., Schulzrinne, H.: Comparison and optimization of packet loss repair methods on VoIP perceived quality under bursty loss. *NOSSDAV'02*, Miami Beach, FL, USA
41. Nasr, M.E., Napoleon, S.A.: On Improving voice quality degraded by packet loss in data networks. the*AFRICON Conference in Africa* 1, 51–55 Sept. (2004)
42. James, J.H., Bing Chen, Garrison, L.: Implementing VoIP: A voice transmission performance progress report. in vol. 42, July 2004, pp. 36-41, (July 2004)
43. Spirent Communications, IPWave, Online Documentation, http://www.spirentcom.com/analysis/product.cfm?WS=13&PR=8
44. Agilent Technologies, Voice Quality Tester (VQT), Online Documentation. http://www.home.agilent.com/USeng/nav/-536885778.536882651/pd.html
45. Sun, L., Ifeachor, E.: Perceived Speech Quality Prediction for Voice over IPbased Networks. in Proceedings of IEEE International Conference on Communications ICC'02, (New York, USA), pp. 2573–2577, (April 2002)
46. Agilent Technologies, VQT Phone Adapter, Technical Specification. http://cp.literature.agilent.com/litweb/pdf/5968-7723E.pdf
47. Ding, L., Goubran, R.A.: Speech quality prediction in VoIP using the xtended E-Model. *IEEE GLOBECOM*, San Francisco, USA, (2003)

48. Conway, A.E., Zhu, Y.: Analyzing voice-over-IP subjective quality as a function of network QoS: A simulation-based methodology and tool. in *Computer Performance Evaluation Modelling Techniques and Tools,Lecture Notes in Computer Science*, vol. 2324. London, UK: Springer, 2002, pp 289-308

49. Avshalom, H.: A Sip of SIP, Lotus Software-IBM SWG, white paper, (Nov. 2003)

50. SIP: Protocol Overview, Radvision white paper, http://www.radvision.com/NR/rdonlyres/51855E82-BD7C-4D9D-AA8A-E822E3F4A81F/0/RADVISIONSIPProtocolOverview.pdf, (2005)

51. Handley, M., Jacobson, V.: SDP: Session description protocol. IETF RFC 2327, (1998)

52. SIP Introduction, http://www.iptel.org/ser/doc/sip_intro/sip_introduction.html

53. Poikselka, M., Niemi, A., Khartabil, H., Mayer, G.: THE IMS: IP Multimedia Concepts and Services, 2nd edn. Wiley, Chichester, West Sussex, England (2006)

54. Handley, M., Schulzrinne, H., Schooler, E., Rosenberg, J.: SIP: session initiation protocol. IETF RFC 2543, March (1999)

55. SIP Server Technical Overview, Radvision white paper, http://www.radvision.com/NR/rdonlyres/0AFA30DF-DAD6-461D-943C-ED33F3E7ABD8/0/SIPServerTechnicalOverviewWhitepaper.pdf, (2004)

56. Roberts, J., Mocci, U., Virtamo, J.: Broadband Network Teletraffic. Final Report of Action COST 242. Springer, Berlin (1996)

57. Park, K., Willinger, W.: Self-Similar Network Traffic and Performance Evaluation. John Wiley & Sons, New York, USA (2000)

58. http://www.tele.dtu.dk/teletraffic/handbook/telehook.pdf, accessed Feb. (2007)

59. Daigle, J., Langford, J.: Models for analysis of packet voice communications systems. *IEEE Journal on Selected Areas in Communications* vol. SAC-4, no. 6, pp 847-55, (Sept. 1986)

60. Kendall, D.G.: Some problems in the theory of queues. *Journal of Royal Statistical Society*, Series B 13(2), 151–173 (1951)

61. Jenq, Y.C.: Approximations for packetized voice traffic in statistical multiplexer, in *Proceedings of INFOCOM'84*

62. Kleinrock, L.: Communication Nets: Stochastic Message Flow and Delay. McGraw-Hill, New York (1964)

63. Østerbø, O.: *Models for End-to-end Delay in Packet Networks Queuing*, R&D report 4/2003

64. Bertsekas, D.P., Gallager, R.: Data Networks. Prentice-Hall, New Jersey, USA (1992)

65. Molina, E.C.: The Theory of probabilities applied to telephone trunking problems. Bell System Tech. J. pp 69-81 (1922)

66. Stallings, W.: High Speed Networks and Internets. Prentice-Hall, New Jersey, USA (2002)

67. Cox, D.R., Isham, V.: Point Processes. Chapman and Hall, Boca Raton, FL (1980)

68. Crommelin, C.D.: Delay probability formulae when the holding times are constant. Post Office Electrical Engineers Journal **25**, 41–50 (1932)

69. Iversen, V.B.: Exact calculation of waiting time distributions in queueing systems with constant holding times. NTS-4, Fourth Nordic Teletraffic Seminar, Helsinki, (1982)

70. Kleinrock, L.: *Queueing Systems*, vol. II: Computer Applications. New York: Wiley, (1976)

71. Wang, L., Verma, P.K.: Impact of Bounded Delays on Resource Consumption in VoIP Networks, submitted to the IASTED International Conference Communication Systems, Networks, and Applications, CSNA 2007, Beijing, China, October. 8-10, (2007)

72. Janssen, J., De Vleeschauwer, D., Buchli, M., Petit, G.H.: Assessing voice quality in packet-based telephony. *IEEE Internet Computing* pp. 48–56, May-June (2002)

73. Clearing the Way for VoIP, An Alternative to Expensive WAN Upgrades, white paper, Gen2 Ventures, (2003)

74. Miyahara, H., Teshigawara, Y., Hasegawa, T.: Delay and throughput evaluation of switching methods in computer communication networks. pp 337–344, (Mar. 1978)

75. Kelly, F.P.: *Blocking Probabilities in Large Circuit-switched Networks*, Adv. Appl. Probab., (1986)

76. Frost, V.S., Melamed, B.: Traffic modeling for telecommunications networks. IEEE Communications Magazine **32**(3), 70–81 (1994)
77. Heffes, H., Lucantoni, D.M.: A Markov modulated characterization of packetized voice and data traffic and related statistical multiplexer performance. *IEEE JSAC*, Sept. 1986, vol. SAC-4, No. 6, pp. 856-868, (Sept. 1986)
78. Cao, J., Cleveland, W., Lin, D., Sun, D.: Internet traffic tends toward Poisson and independent as the load increases. In: Holmes, C., Denison, D., Hansen, M., Yu, B., Mallick, B. (eds.) Nonlinear Estimation and Classification. Springer, New York (2002)
79. Sharafeddine, S., Riedl, A., Glasmann, J., Totzke, J.: On traffic characteristics and bandwidth requirements of voice over IP applications. in *Proc. 8th IEEE International Symposium on Computers and Communications (ISCC)*, pp. 1324–1330, (2003)
80. Gardner, M.T., Frost, V.S., Petr, D.W.: Using optimization to achieve efficient quality of service in voice over IP networks. Proc. Int. Conf. Performance, Computing, and Communications, Phoenix,Arizona, April 2003, pp. 475-480
81. Kleinrock, L.: *Queueing Systems,*vol. I: Theory. New York: John Wiley and Sons: (1975)
82. Verma, P.K.: Performance Estimation of Computer Communication Networks: A Structured Approach. Computer Science Press, Rockville, MD (1989)
83. Harada, H., Prasad, R.: Simulation and Software Radio for Mobile Communication. Artech House, Boston, MA (2002)
84. Wang, L., Tachwali, Y., Verma, P.K., Ghosh, A.K.: Impact of bounded delay on throughput in ulti-hop wireless sensor networks. *Journal of Network and Computer Applications*, Elsevier, March. (2009)
85. Van Der Wal, K., Mandjes, M., Bastiaansen, H.: Delay performance analysis of the new Internet services with guaranteed QoS. *Proc. IEEE* 85(12), 1947-1957 (1997)
86. Tobagi, F.A.: Modeling and performance analysis of multihop packet radio networks. *Proceedings of the IEEE* 75(1), 135-155 (Jan. 1987)
87. Mandjes, M., van der Wal, K., Kooij, R., Bastiaansen, H.: End-to-end delay models for interactive services on a large-scale IP network, Seventh IFIP workshop on Performance Modelling and Evaluation of ATM Networks: IFIP ATM'99, Antwerp, Belgium, June 28-30, (1999)
88. De Vleeschauwer, D., Petit, G.H., Steyaert, B., Wittevrongel, S., Bruneel, H.: Calculation of end-to-end delay quantile in network of M/G/1 queues. *IEEE Letters* 37, 8, (12), 535–536 (2001)
89. Shortle, John F., Brill, Percy H.: Analytical Distribution of Waiting Time in the M/{iD}/1 Queue. Publication??? 50, 185-197 (July 2005)
90. Ramamurthy, G., Sengupta, B.: Delay analysis of a packet voice multiplexer by the $\sum D_i/D/1$ queue. *IEEE Transactions on Communications* 39(7), 1107-1114, July (1991)
91. Østerbø, O.: An approximative method to calculate the distribution of end-to-end delay in packet network. R&D Report, (Feb. 2002)
92. Østerbø, O.: End-to-end queuing models with riority. R&D Report (Mar. 2003)
93. Iversen, V.B., Staalhagen, L.: Waiting time distribution in M/D/1 queuing systems. Electronics Lett. **35**, 2184–2185 (1999)
94. Takagi, H.: Queueing Analysis, Volume 1: Vacation and Priority Systems, Part 1. Amsterdam, North-Holland, (1991)
95. Gopal, P.M., Kadaba, B.: A Simulation study of network delay for packetized voice. in *Proceedings of GLOBECOM'86*, Dec. (1986)
96. Hancock, J.: Jitter Fundamentals. *High Frequency Electronics*, pp?, April (2004)
97. Jitter Solutions for Telecom, Enterprise, and Digital Designs, Agilent Technologies, August (2005)
98. Davies, N., Holyer, J., Thompson, P.: End-to-end management of mixed applications across networks. in *IEEE Workshop on Internet Applications*, pp. 12-19, (1999)
99. Karol, M., Krishnan, P., Li, J.J.: enProtect: enterprise-based network protection and performance improvement, Avaya Labs Research-Technical Report, August (2002)

100. Guilleman, F., Roberts, J.W.: Jitter and bandwidth enforcement. in*Proc. IEEE Globecom '91*, Phoenix, AZ, pp. 261-265

101. Wang, L., Verma, P.K.: Impact of bounded delays on resource consumption in VoIP networks. in IASTED International Conference Communication Systems, Networks, and Applications, CSNA, (Oct. 2007)

102. Dahshan, M.H., Verma, P.K.: Resource based pricing framework for integrated service networks. Journal of Networks 2, 36–45 (2007)

103. Wang, L., Verma, P.K.: The Notion of cost and quality in packet switched networks an abstract approach. in *IEEE CQR 2009 International Workshop*, Naples, Florida, May 12-14, (2009)

104. Black, U.: Voice over IP. Pretice Hall, New Jersey, USA (1999)

105. Johnson, C.R., Kogan, Y., Levy, Y., Saheban, F., Tarapore, P.: Voice reliability: A service provider's perspective. *IEEE Commun. Mag.* pp. 48-54, (July 2004)

106. TSB116: Voice quality recommendations for IP telephony (2001)

107. DaSilva, L.A.: Pricing for QoS-enabled networks: a survey. *IEEE Communications Surveys & Tutorials* 3(2), 14-20

108. Falkner, M., Devetsikiotis, M., Lambadaris, I.: An overview of pricing concepts for broadband IP networks. *IEEE Communications Surveys & Tutorials* 3(2), 2-13

109. Zhen, L., Wynter, L., Xia, C.: Usage-based versus flat pricing for E-Business services with differentiated QoS. *IEEE International Conference on E-Commerce-CEO '03*, pp 355-362

110. Fishburn, P.C., Odlyzko, A.M.: Dynamic behavior of differential pricing and quality of service options for the internet. 1st International Conference on Information and Computation Economies, Charleston, South Carolina, USA, (1998)

111. Wang, L., Verma, P.K.: Cumulative impact of inhomogeneous channels on risk, The 6th International Symposium on Risk Management and Cyber-Informatics: RMCI 2009, Orlando, Florida, July 10-13, (2009)

112. Harrald, John R., Stephens, Hugh W., vanDorp, Johann Rene: A Framework for sustainable port security. *Journal of Homeland Security and Emergency Management* 1(2), (2004)

113. Stephens, H.W.: Barriers to Port Security. *Journal of Security Administration* 12(2), 29-40

114. Flynn, Stephen: The Edge of Disaster. Random House, New York (2007)

115. Lewis, J.M., Lakshmivarahan, S., Dhall, S.K.: Dynamic Data Assimilation: A Least Squares Approach. Cambridge University Press, Cambridge (2006)

116. Kivman, G.A.: Sequential parameter estimation for stochastic systems. Nonlinear Processors in Geophysics 10, 253–259 (2003)

117. Nussbaumer, H.J.: Fast Fourier Transform and Convolution Algorithms. Springer-Verlag, Berlin Heidelberg (1990)

118. Grosswald, E., Kotz, Samuel: An integrated lack of memory characterization of the exponential distribution. Annals of the Institute of Statistical Mathematics 33, 205–214 (1981)

119. Shimizu, K., Crow, E.L.: Lognormal Distributions: Theory and Applications. Marcel Dekker, INC, New York (1988)

120. Earthquake Facts and Statistics, USGS, http://neic.usgs.gov/neis/eqlists/eqstats.html

121. Bob Briscoe, Andrew Odlyzko, and Benjamin Tilly: Metcalfe's law is wrong, *IEEE Spectrum*, July (2006)

122. Yingzhen Qu and Pramode Verma: IEEE Communication Letters also referenced elsewhere in the book

123. Gillet, Sharon., Vogelsang., Ingo (eds.): Competition, Regulation, and Convergence, p. 310. Lawrence Erlbaum Associates, New Jersey (1998)

124. Wang, L., Verma, P.K.: A Network based authentication scheme for VoIP. *IEEE International Conference for Communication Technology ICCT'06*, Guilin, China

125. John Franks, Phillip Hallam-Baker, Jeffrey Hostetler, Scott Lawrence, Paul Leach, Ari Luotonen, Lawrence Stewart: HTTP authentication: Basic and digest access a uthentication, IETF RFC 2617, (June 1999)

126. Kent, S., Atkinson, R.: IP Encapsulating Security Payload (ESP), RFC 2406, Nov. (1998)
127. Anwar, Z., Yurcik, W., Johnson, R.E., Hafiz, M., Campbell, R.H.: Multiple Design Patterns for Voice over IP (VoIP) Security, in*Proc. 25th IEEE International Performance Computing and Communications Conference (IPCCC)*, Phoenix, Arizona, USA, pp. 485-492, April. 10-12, (2006)
128. Dierks, T., Allen, C.: The TLS Protocol Version 1.0. IETF RFC 2246, January (1999)
129. Housley, R., Polk, W., Ford, W.: InternetX.509 Public Key Infrastructure: Certificate and CRL Profile. IETF RFC 3280, April (2002)
130. Salsano, S., Veltri, L., Papalilo, D.: SIP security issues: The SIP authentication procedure and its processing load. *IEEE Network* 16(6), 38–44, Nov.–Dec. (2002)
131. Ramsdell, B.: SMIME Version 3 Message Specification. IETF RFC 2633, June (1999)
132. Orrblad, J. Alternatives to MIKEY/SRTP to secure VoIP. Master Thesis, KTH Royal Institute of Technology, Stockholm, March (2005)
133. Bilien, J., Eliasson, E., Orrblad, J., Vatn, J-O.: Secure VoIP: call establishment and media protection. 2nd Workshop on Securing Voice over IP, Washington DC, June (2005)
134. Ono, K., Tachimoto, S.: SIP signaling security for end-to-end communication. in. *Proc. 9th IEEE Asia-Pacific Conf. Commun.*, Penang, Malaysia, pp. 1042–1046, (2003)
135. Stallings, W.: Cryptology and Network Security. Prentice-Hall, Englewood Cliffs, NJ (2005)

Index